The Quantum Kinetic Well

The Quantum Kinetic Well

Powering the World with Endless Clean Energy

Riley D. Lee

McKane B. Lee

Quantum Kinetics Corporation

CONTENTS

1

The Story

The Quantum Kinetic Well^{TM}

Powering the World with Endless Clean Energy

Written by: Riley D. Lee

2

Dedication

"I dedicate this book to my father - my best friend."

--McKane B. Lee

3

About the Author

The inventor's favorite fictional character is Iron Man, Tony Stark, first introduced to the world by Marvel in 1963. Comic book lore depicted Tony's father as a notable inventor who, inspired by the idea and hope of sustainable fusion power, developed an electromagnetic machine that could provide limitless clean renewable energy to the world. Although the project ultimately failed to deliver the promise of fusion power on a large scale, its underlying technology was used by Tony Stark to build a modular version of a fusion power plant – the Arc Reactor, the source of Iron Man's secret powers.

Over 50 years of superhero comic book tales and blockbuster movies, the saga of Iron Man's Arc Reactor has inspired many real-life inventors to follow the same dream of fusion power for the betterment of mankind. Recently, MIT has been working on a reactor that would function like the arc reactor, although the science team avoids using the same name. Other nuclear fusion projects currently under development around the world also pursue the same goals by heating matter to more than a hundred million degrees until the super-heated atoms melt into a magnetically confined plasma field, thus replicating conditions in the center of a star. This is the "star building" model to produce nuclear fusion. Despite investments of billions of dollars into research and development, the goal of producing sustainable stellar-like fusion remains as yet unsolved due to overwhelming complications and technical challenges.

Quantum Kinetics Corporation's ("QKC") mission is the realization of the idea and hope of sustainable fusion power on a universal scale. QKC's revolutionary technologies are based on the discoveries of one Inventor who, inspired by science fiction and his own imagination, sought to address the lofty goal of developing clean energy machines similar-to the "star building" projects, without resorting to extreme high temperatures and super-sized molecular colliders. This project is the formal introduction of that novel, renewable energy platform to the world – and for the first time in history – realization toward a functioning Arc Reactor and Quantum Thruster.

Truth is stranger than fiction...

Repulsor Technology - Water mist (7uL) instantly turned to 'Thermal Explosive Energy' (GTNT) - Star-Burst. Milli-ampere energy input.

4

Chapter 1: A Fateful Road Trip

McKane remembers Happy. She regularly scolded him for spending too much time on his experimental, hydrogen-powered car out in the garage.

It's not going to work.

Your BMW engine is just fine the way it is.

It gets good mileage now.

What are you running from?

Can you come out of the garage for a minute?

He recalls one of his dad's quips about the two secrets to a long marriage. One is learning to say "Yes dear," and the other is making enough money to keep her in the style to which she is accustomed.

But McKane is a firefighter. He isn't planning on making a fortune any time soon. So, he'll have to content himself with working on the first secret.

"Yes, I'll be right in, honey."

He and Happy were friends in high school but never officially dated. Happy grew up in a communal setting on a 291-acre ranch shared by 350 members of the Love Israel Family. It was a hippie community born out of the counter-culture movement in the 1960s. Happy's father, a clinical psychologist, had been part of the academic arm of the movement in California. His associates were famed psychotherapists and authors, Carl Rogers, and Fritz Perls.

Happy grew up accepting that she had 100 brothers and sisters on the ranch. Her bedroom was a small, unfinished cabin with a dirt floor. Every summer, the cabin was invaded by ladybugs, the source of her uneasiness around insects. She got involved in gourmet cooking early. The food supplied small eateries run by the family in Snohomish County. She also fashioned wonderful crafts to be sold at the annual Garlic Festival, attended by thousands.

He recalls having lunch with Happy in the high school cafeteria just before their graduation. Present as usual were his coterie of female admirers sitting too closely together at the round table. Happy manages to slip in next to him. She gets his attention long enough to whisper in his ear, "Let's meet again back here when we are both 30 and get married." He agrees to the proposed rendezvous, thinking it impossible to be single in twelve years.

They have been married for two years now. They live in a small rental house on a road with an Indian name: Sioux Drive. Family lore has McKane inheriting Indian blood on his mother's side.

He works at the fire station in Stanwood. She works in Bellingham as a food scientist for a large nutrition company.

They are getting along and still in love. He is thinking it is too soon for children.

* * *

Happy is planning a weekend road trip to Leavenworth. McKane will come along, even though he doesn't like driving in the snow. She hopes Alex and Kaitlin will come, too. Alex works with McKane at the fire station. Kaitlin works in nutritional health at Skagit Hospital. The two couples have everything in common.

Happy plans for the group to celebrate Alex's birthday on their brief mountain holiday. She is thinking they could arrive in Leavenworth after work on Friday night and catch a hotel room. The next day would be open for tourist activities and for viewing the Christmas lights, still up after the holiday season. She can drive her Subaru. It has four-wheel drive.

No, she doesn't need snow tires for the Subaru.

The sun is shining on a frosty Saturday morning in Leavenworth. McKane, Happy, Alex and Kaitlin are following a snow-covered sidewalk from the hotel into town. The respective couples walk, arm in arm, in the frigid air. They notice gift shops, craft stores and small eateries along the way.

They stop at an outdoor beer garden and order the obligatory stein of beer. Next, they pass by *A Matter of Taste*, a Bavarian gift shop and cafe. McKane and Alex walk in. They notice a taunting sign of warning on the front counter next to a hot sauce bowl: *Hotter Than Hell. Try at Your Own Risk.*

They accept the challenge. McKane is first to take the plunge into the hot sauce with pretzel in hand. Alex, not to be outdone, follows his buddy. A moment later they are dashing across the street to the Gelato shop for oral first aid.

In the evening the couples find an Italian restaurant. They order too much pasta to finish. In the waning sunlight they linger at the table talking about the day. The evening continues with a walk along a boulevard framed by festive lights of the season. They stop in at an old-fashioned, Bavarian drinking establishment where they enjoy sample-sized glasses of beer from local breweries. Before long, the platter is full of empty glasses.

A young man approaches. He gives Happy a big bear hug, then turns to introduce himself to the others. "Hello, I'm Sam, an old friend of Happy's family." Sam lives nearby, not far from Leavenworth, and at a higher elevation. Sam officially invites them to Sunday morning brunch on the mountain.

Happy accepts without consulting the others.

* * *

McKane wakes early the next morning. He hadn't slept well. He wonders if it was the alcohol or a frightening dream. Maybe it's both.

He presses to recall the dream, sensing it may contain an embedded message, perhaps an insight or even a warning. His search only reveals darkness.

Darkness, swallowed by a deep abyss.

Unsettled by the disturbing vision, he tightly closes his eyes and draws the covers over his head wishing to escape the looming black hole.

The others are still sleeping.

He thinks about their original itinerary for the day: check out at 11 a.m. and then a two-hour drive back home through the mountain pass. Happy's phone lights up—a text. It is Sam confirming the invitation. She replies, "Yes." McKane is startled by the unilateral change to the group's initial plan.

It is growing colder in Leavenworth. Outside, light snowflakes are beginning to fall.

McKane checks the weather forecast for Stevens Pass. He opens the weather app on his cell phone. The forecast calls for heavy snow and winter storm warnings. He rolls over to tell Happy and sends out a text message to Alex and Kaitlin. He argues that with the approaching weather system it is safer to head west and home rather than east to higher altitudes. Happy insists that she is an experienced driver in the snow. She adds that her friends are expecting them. Not wanting to disappoint Happy or Sam, the others agree.

An hour later, Happy pauses the Subaru at the base of the ridge road leading to the ranch house.

It is an old logging road, converted to community access. It looks rough and steep. There are sharp switchbacks every few hundred feet. The air is even colder here than in Leavenworth. The wind is stiffer. The snow is still falling.

Heavy snow.

McKane says they should just turn back. Alex and Kaitlin are looking at him nervously. Happy is concerned, too. She texts messages Sam. He answers, offering to pick them up in his retro-fitted, four-wheel-drive truck equipped with chains, sandbags, snow tires, and a snowplow.

But she doesn't convey Sam's offer to the others.

Happy's Subaru ascends the ridge road through swirling snow. As it moves upward, the SUV encounters an ever-steepening slope. The slate gray skies are darkening by the moment. McKane notices accumulations of snow building on the road. The Subaru is doing well even without snow tires or chains. They will probably make it. The lodge is near the top of the ridge.

They arrive unscathed.

Greetings and introductions buzz with energy. McKane feels distracted. When it is possible, he escapes the happy chatter and meanders into the lodge's great room. He finds himself facing windows that reveal panoramic views of the valley below. He watches distinctly layered cloud formations approaching over the adjacent peak. The wind pulsates the windows. The front is moving in fast.

Soon, a blizzard encircles the house.

Given their uncertain arrival time, preparations for brunch haven't begun. And of course, they had to catch up on current events first. The well-planned gourmet brunch will be late. The chef of the outfit, Happy takes charge of making a delicious quiche.

McKane notices again the drifting snow.

After brunch, McKane contrives excuses for announcing their early departure.

Meeting you has been wonderful.
We would love to stay longer, but maybe next time.
We all work tomorrow.
We really need to get on the road.

* * *

The descent down the ridge road seems to take forever. McKane and Alex are positioned in the back seats. Kaitlin is the front seat passenger. Happy drives cautiously. After all, she is winding through an 8,000-foot mountain range after a heavy snowfall in a small SUV without chains or snow tires.

As the downslope steepens, Happy manages to keep the car's momentum under control by gently pumping the brakes. She is even more deliberate at the sharp switchbacks. The Subaru is moving in slow motion when it encounters an impossible turn on the icy corridor.

Suddenly, the SUV is flying, catapulted into a white abyss.

The car careens hundreds of feet down the steep landscape. Kaitlin begins screaming when the car severs the first tree, then another.

The car somersaults.

Glass is shattering.

Metal is creaking, loudly.

The engine is racing out of control. Untethered objects are airborne within the chamber.

McKane remembers the SUV's roof getting smaller and smaller with each roll-over. They are trapped. *Is this how it is going to end?*

Maybe this is a nightmare.

He tries to wake up, but his effort only confirms the terrible truth.

This is real. Our lives will soon be over.

He pushes against the collapsing roof with all his strength. The Subaru is still shrinking. He counts the roll-overs until it gets to twelve. *On the thirteenth, we will all be dead*. The car suddenly stops upside down against a tree, quiet and motionless in the cold mountain air.

The resting spot...

Somehow, he is alive. So are the others.

* * *

McKane exits the car, blood streaming down his forehead. His face is throbbing, beaten hard by loose, flying objects in the car. Happy's jars of canned vegetables, homemade jams, and jellies. His back, shoulder, and neck send agonizing messages of concern.

Despite the injuries incurred, his training kicks in, directing decisive action in the moment of danger. His full focus is rescue.

McKane makes his way to the others still trapped within the mangled wreckage. Quickly, he checks medical status to see if they can be safely moved out of the vehicle. First, he evacuates Happy and Alex from the devastation. Kaitlin cannot stand up. Her leg is badly injured.

Alex and McKane work together to pull her out of the vehicle.

Then McKane calls 911.

* * *

The local firefighters arrive on the scene. McKane asks if they are ready to perform a "high angle rescue."

They cannot comprehend the question. "What is a high angle rescue?" they ask.

"We are only volunteers," they explain apologetically.

Stunned by the admission, McKane realizes there is no option but to take command of the situation. He directs the volunteer firefighters to get the all the rope out of the rig, plus the spider straps and backboard.

"We don't know where the equipment is stored," whine the volunteers.

He commands them to find it anyway, and fast.

At the edge of the precipice, he directs the volunteers to tie a Bowline knot on the end of the backboard.

"What is a Bowline knot?"

McKane, now the teacher, shows them. One end of the rope is tied to a tree and the other to the backboard. Quickly now, the backboard is lowered to the injured parties below. Kaitlin is most injured. Alex and McKane wrap her in a blanket and carefully secure her to the backboard. Together, they hoist her to the ambulance above.

The paramedics arrive just as Kaitlin is placed in the ambulance. Then the sheriff arrives.

"You should all be dead," he says matter-of-factly.

* * *

As the ambulance is about to begin the transport, McKane stands at the open back door and watches Kaitlin and Alex. He sees Alex's impassioned look towards Kaitlin on the stretcher. He notices her ardent and tearful gaze back to Alex. Their reciprocal messages—wordless but undeniable.

McKane slowly but firmly shuts the ambulance doors, realizing that their life together is just beginning while a precious part of his own life is ending.

* * *

He catches a ride to the hospital to have the lacerations on his face, head and arms treated. The shoulder and back injuries can be treated later when he returns to the west side. He and his friends cheated death.

He should be happy, but he is feeling sad and confused. He wants to cry. He won't cry, certain of a deeper descent into the swirling abyss.

He calls his parents from the hospital and asks for a ride home. He explains the accident. He and Happy are okay, but her car is destroyed.

They both need to work on Monday.

On the return trip, there is silence. Silence, and a growing tension between them.

5

Chapter 2: A New Direction

His job is firefighting. He has been at it for over two years if you count his training at the Academy.

You go on eight to fifteen calls a day. And each of those calls is the worst day of someone's life. It could be a drowning, heart attack, motor vehicle accident, electrocution, amputation, degloving incident or a fire. He learns how to emotionally detach just to get through the day.

McKane tries to shake off the accident. It's time to return to normal life. That thinking works up until the moment the flashbacks start.

Images in his mind haunt him. The flashbacks worsen whenever he sees an emergency response vehicle with activated lights and sirens. He talks to friends about how nervous he feels in the backseat of a vehicle. His friends are kind and sympathetic, but they can't know how terrible the accident was.

The flashbacks make it hard for him to work at the fire station. He begins to pass off his regularly scheduled hours to other firefighters who want the time. He mentions to the battalion chief that the accident had been intense. He's not doing well. The battalion chief listens to McKane, as he recounts the flashbacks and nightmares.

One dream involves taking an emergency call, arriving on the scene of a horrific accident, and then seeing himself dead in a ditch. He tells the chief he relives the accident whenever the station bell goes off. He fears that he'll make a mistake that could hurt someone. Safety is the heartbeat of the job—he can't afford to compromise anyone's safety.

The battalion chief is understanding and gives him one month off. The chief encourages him to get better.

"You are a good firefighter. We don't want to lose you."

One month later, McKane returns all his gear to the station.

* * *

McKane needs to replace lost income, so he turns his attention to building up the business he started a few years back, a pole-vaulting club called *Pro Vault NW*. The club occupies part of a large gymnastics' facility near the Boeing plant in Everett. He owns all the poles, pit, standards, runway, small tools, computers, and video equipment. There is a big whiteboard for teaching sessions, leaving messages, or doodling.

The facility is as good as or better than any like it in the country.

He teaches the pole vault with his own unique methods. His approach is novel and 180-degrees out of sync with traditional coaching. McKane invents teaching aids and builds novel training equipment to train budding athletes. Athletes from Texas to British Columbia fly in to ProVault to learn the skills he mastered while in high school.

As a high school freshman McKane's coaches first recognized his vaulting talent. They projected he would be state champion. To them he looked like a smaller version of Sergei Bubka, in the air but not on the runway. He glided, not jetted down the runway.

McKane set a state record of fifteen feet nine inches when only a sophomore. The next year he jumped sixteen feet eight and a half inches, easily winning the state championship. It helped that McKane had a background in gymnastics. Before pole vaulting, he had twice been all-around state gymnastics champion.

In McKane's vision, pole vaulting was just gymnastics on the end of a stick.

The clients at ProVault include athletes from age 12 to 82. They soon grow into a family of dedicated athletes, making the training sessions a blast. Sometimes a pizza party breaks out. In the summer, the group spends entire days pole vaulting in the sand along the Puget Sound. Crowds of spectators gather around loudly cheering successful jumps. Some of ProVault's clients enter national championships and win. Others earn college scholarships.

To McKane's satisfaction, his athletes begin to look up to him. They like him not just because he is an excellent teacher, but because he seeks to be every trainee's friend.

Some of the athletes develop closer ties to McKane. Greg, a high school senior with aptitude in many sports, is one such athlete. A football and soccer star, success on the field comes easy to Greg. His high school coach predicts he will be drafted into the NFL as a field goal specialist or maybe a fast linebacker. At the beginning of his senior year, Greg expands his athletic interests to include the pole vault. Within eight months of training sessions at ProVault, Greg is vaulting at big bars that should win him the title at the state meet.

Three weeks before the championship meet, however, Greg incurs injuries in a serious accident. A vehicle crashes into his car from behind. Greg's legs and back are aching, but he drags himself to vault practice. He finds he can't get back up on the big poles. Nevertheless, Greg competes at the state meet, playing through lingering injuries. He places a disappointing fourth. Not only is Greg's brief vaulting career over, but his entire athletic future disappears.

Greg enrolls in a university in California without the aid of an athletic scholarship, but he keeps in touch with McKane. Greg is hoping his vaulting coach can still serve as his life coach—and lifelong friend.

* * *

Eighty-two-year-old Butler is a "butler" in the true sense of the word, doing everything he can to serve everybody around him. Butler owns a company that builds parts for Boeing and for military aircraft. At age 80, he tries to retire but his wife won't let him. At age 82, he is still working.

McKane predicts he will probably die with his boots on.

Butler develops an exceptionally strong bond to McKane. They first meet in the summer at an "All-Comers" track meet in Edmonds. They share a birthday, December 4th.

Butler and McKane during a lunch break at the machine shop

Curiously, it is also the birthday of Sergei Bubka, the world's most famous pole vaulter—the athlete who broke the men's world record 35 times. Both McKane and Butler had learned to vault in their own backyards as children. Butler vaulted in high school and then later in the military. McKane vaulted in high school and then as a scholarship athlete at the University of Washington. Beyond a common birthday, they share a lifelong passion for the sport.

As their friendship grows, Butler discovers that McKane lost his four grandparents early in life. It is a loss McKane feels mostly around holidays. Butler being "butler" decides to appoint himself as McKane's honorary grandfather.

Thereafter, they make sure to meet or talk by phone every day.

It becomes habit for McKane to drop off lunch to Butler at the company shop, partly out of concern that Butler isn't eating enough. These shared lunches give the two of them structured time to talk about everything. Butler learns of McKane's secret dream to build a water-powered car. Butler relates to McKane's passion for invention, having developed hydrogen and oxygen splitting devices himself. In fact, he has been at it for a lifetime.

Having become acquainted with McKane's aspirations as an inventor, Butler allows McKane direct access to his plant and equipment. There McKane continues his work, taking advantage of the material inventory in Butler's plant. He learns to use the sophisticated machines and even assists the crew at the shop running the lathe.

Involving McKane in the production process is Butler's way of encouraging his surrogate grandson to follow his dreams.

* * *

At officially sanctioned track meets, McKane interacts with other vault coaches employed at high schools across the state, but never feels welcome or part of the coterie of school coaches.

He's a private coach—a businessman.

At the stadiums, the other coaches try to keep him off the field so he can't interact with the athletes. They make threats to call the police and have him arrested for trespassing on public property.

Some of the more aggressive coaches openly insult him in front of the athletes. One coach calls McKane a "piss ant" as he extends his hand in a mock friendly gesture at the state meet. McKane winces but keeps his head high in front of his athletes.

He's resolved not to surrender to the united opposition against him. He remains a constant presence at the track meets—and his athletes love him all the more for it.

* * *

Butler's favorite aphorism is "In the pole vault, an athlete can look so good or so bad and the difference is so little." Small differences in technique make all the difference. McKane uses the big white board at his facility to teach the science and the precision of the pole vault.

McKane's background in art helps him configure precise diagrams and illustrations. As a younger man, he had earned a master's degree in 3D Digital Art from the University of British Columbia, Simon Fraser University, Emily Carr University of Art and Design, and the British Columbia Institute of Technology. He also taught the Unity 3D Game Engine and Digital Art in Europe and Washington state for four years. For this reason, he feels comfortable and confident in front of a classroom, even without formal training in education.

Most days, the ProVault white board is covered with intriguing sketches with arrows, vector lines, symbols, and numbers. McKane comes to realize that even as a vault coach, he thinks like a mathematician—and an artist.

Sometimes after practice, McKane doodles at the white board, drawing illustrations of a theoretical hydrogen injector for his old BMW. His idea is to find a way to gain free energy with a new water splitting technique. McKane devises a simple, but novel electronic system to activate the release of hydrogen and draws it up on the white board.

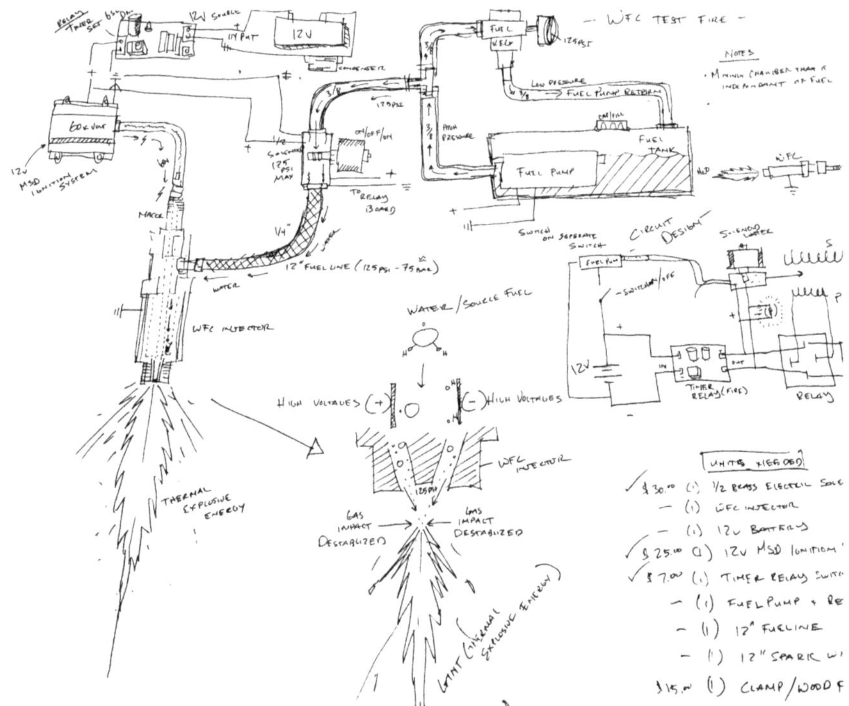

McKane's drawing of the hydrogen injector system

On a few occasions, he leaves the after-hour sketches of his inventions. At the sight of them the athletes are fascinated and barrage McKane with questions about his theories. For no charge, he offers them a layman's view of science, invention, spacecraft, clean energy, and fusion. McKane notices that the younger the athlete the more he or she intuitively understands the science he presents.

Nevertheless, some of the students' parents think he is a little crazy.

* * *

Life at ProVault comes to delight and comfort McKane. In his facility a group of wildly diverse athletes evolve into a loving, supportive community of friends. McKane embraces his organic role as their leader and enjoys his athletes' devotion and respect.

Nevertheless, in a close-knit circle of friends, it is hard to keep a secret.

Everyone at ProVault knows about McKane's accident—and its steep cost for him. Knowing his concern and dedication to each of them, his athletes are sympathetic and understanding in return.

Some even offer him hugs when he appears down.

After the accident athletes comfort McKane

6

Chapter 3: A Necessary Detour

Battling a bout of depression for the first time in his life is difficult. It comes with a puzzling sadness that permeates everything. Losing himself in dysphoria feels like losing a portion of his soul, only worse.

McKane realizes he needs help to fully recover in the wake of his accident, and to reclaim what he has lost emotionally. His father, a clinical psychologist, refers McKane to a neuropsychologist famed for working with patients suffering with Post Traumatic Stress Disorder.

Dr. Burkhart's office is in Edmonds, a sprawling waterfront community with timber town roots. A Washington state ferry terminal is the city's beating heart. It is a gateway to Whidbey Island and the Olympic Mountains to the west, and the San Juan Islands and British Columbia to the north. The waterfront is lined with cafés, coffee shops, art galleries, a museum, a public marina, and a city park.

Edmonds is a beautiful city. The only problem is that the drive into Edmonds can sometimes be frustrating. Ferry rush hours feature congestion that rival I-5 traffic jams, long backups, and impatient commuters.

The first visit to Dr. Burkhart's office catches McKane off guard. He doesn't see the doctor as expected. Instead, the doctor's trusted assistant,

Lee Farley, M.A., meets him there. Ms. Farley explains that McKane must undergo some psychological tests before seeing Dr. Burkhart.

McKane wonders, *what is a psychological test*? *Will I pass*? He concludes, *this is going to be a very serious, professional process*.

Ms. Farley asks him why he wants to see the doctor in the first place. In response McKane simply relates his memories of a car accident.

"I am a backseat passenger in a SUV. We are descending a steep mountain road. There is a snowstorm and a cliff . . . suddenly, we are flying. The car lands and begins tumbling. The car's roof is shrinking with each somersault. I push against the roof. It keeps buckling. Windows are crunching. My friends are screaming . . . I know we are going to die. I count the rollovers until the number twelve. I think, *this can't be happening*. *It must be a bad dream*. We survive. Now I'm here."

He mentions to her the flashbacks—how he's losing interest in golf and pole vaulting. How he feels nervous in the back seat of a car. How he's lost a little weight and his appetite.

How he's getting a divorce.

How he feels like a total failure.

McKane's had recurring nightmares about a ghost—a ghost who reminds him that he almost died without accomplishing his special purpose.

What that purpose is, the ghost doesn't tell him.

* * *

The first test is written. McKane reads the instructions. The page looks like a checklist. You just check a box if a listed problem is true for you. Sitting in the small interview room alone, he almost involuntarily checks the boxes "Depressed" and "Anxious."

He thinks, *what the hell am I doing with depression and anxiety*?

In the next test, McKane moves around puzzle pieces in a game called "Tower of Hanoi." Ms. Farley explains the task is to reorganize eight sizes of wooden disks, stacked largest to smallest in pyramid fashion, from a single peg on one side of the puzzle to a single peg on the opposite side of the puzzle. There is a central peg the player can use to accomplish the task as needed. Ms. Farley makes clear there are only two rules in the game. One is that only one disk can be moved at a time. The second rule is that a larger disk can never be placed on top of a smaller disk.

What Ms. Farley does not explain is that it takes 255 moves to correctly solve the puzzle.

McKane gets off to a good start. He easily moves five disks from one side of the puzzle to the other by shuttling the disks on and off the central peg. It's a good strategy. He might solve the puzzle in record time. Suddenly, he is stuck. He can't make another move without breaking Ms. Farley's rules.

His breathing tightens.

He senses the walls of the small exam room are shrinking. Thoughts embedded in panic accelerate. He feels frozen in space and time, trapped inside the puzzle. He is replicating his emotions in the collapsing Subaru.

Ms. Farley says he should stop for the day.

* * *

The first meeting with Dr. Burkhart does not go as expected. Dr. Burkhart presents as a normal guy, low-key and laid-back. He is anything but stuffy and over-intellectual, and his informal style sets McKane at ease. The two of them are simply talking about the accident when the patient feels his defenses dropping.

Dr. Burkhart engages in talk therapy first. McKane talks. The doctor listens. Sometimes it is like a casual conversation with a close friend over morning coffee. Other times the dialogue turns more ominous—especially when Dr. Burkhart poses a piercing question.

McKane looks forward to the sessions. He feels better after talking about his work and things at home. He notices how Dr. Burkhart keeps circling conversations back to the accident itself. "What do you remember about the accident?" "What were your feelings right after?"

McKane feels blocked. He can describe the mechanics of the rollovers but not his feelings immediately after the incident. Dr. Burkhart gently but firmly persists. Calmly he repeats the questions, hoping to clinically evoke insight surrounding a latent memory. Instead, the doctor's questions only produce in McKane a rising anxiety.

"I must have been too busy with the rescue to notice," is McKane's half-hearted excuse for his lost emotional memory.

The doctor is not convinced. He senses his patient is still avoiding something—

Something important.

Something deeper.

Dr. Burkhart tries an experiment. McKane is sitting in a chair in front of a row of red lights. The lights are pulsing left to right. He stares at the lights without moving his head. When the lights stop moving, Dr. Burkhart directs him to close his eyes and remember the incident. The process repeats, each time with faster moving lights. On the fifth trial, Dr. Burkhart asks him to close his eyes and see himself at the crash site.

"What are you seeing?"

"I am at the edge of the mountain looking down as the car is flipping over and over. There is a ghost next to me watching. The ghost sees through the trees and through the snow. The ghost is in the trees and in the snow, watching. The ghost turns to whisper in my ear.

It's not your fault.

As these words exit his lips, McKane starts to weep. He doesn't know if the message is for him or Happy. It is probably for both.

"You are certainly taking to this well," Dr. Burkhart gently says.

* * *

McKane continues in therapy for more than a year. Sessions are every Thursday afternoon. Sometimes he forgets that he has severe PTSD. Dr. Burkhart doesn't remind him.

McKane comes to understand that he's not undergoing classical psychoanalysis. Dr. Burkhart employs a unique approach to disentangling depressive patients from past emotional fixations. It's a positive process, gently opening the patient's insight to resolve hidden, intrapsychic conflicts.

McKane comes to accept that his past cannot be changed. And with time, he finds himself feeling more upbeat.

* * *

With one milestone achieved, Dr. Burkhart turns therapy in the direction of McKane's future. Like mending a fractured but invaluable work of art, the doctor understands the importance of post-traumatic growth as an essential factor in stabilizing recovery and finding one's "second life." In his view, successful psychotherapy propels the patient into a bright future built on the ruins of a broken history.

Dr. Burkhart encourages McKane to talk about his hobbies and what it is he would love to be doing for work five years from now.

"What is this thing you are spending all your time on at home?"

McKane explains that he is attempting to develop a water-powered car. Though progress is slow, McKane's eventual success is not out of reach. His principal challenge to success is a lack of time. He wishes his work on the car were his job instead of a weekend hobby. During the conversation, McKane confesses a desperate, if not irrational, urge to return to his hometown to finish his hydrogen-fueled car.

Having said it out loud, McKane is suddenly struck by an absolute conviction of what he must do next.

7

Chapter 4: Into a Secret Forest

Six years in a garage.

The sentence is self-imposed. McKane chose to enter a virtual prison of four high walls and empty space, leaving everything else outside: wife, family, friends, vocation, and all avocations.

McKane is convinced his plan merits the weight of the sacrifice attached to his work. He's no longer a hobbyist. His aim is serious and significant—to develop a synthetic mechanism to trigger a water-to-energy conversion system. Based on an underpinning design he has been grinding out, his objective seems completely possible.

At least theoretically.

If he succeeds to plan, such a device would generate an earthquake in the world of energy production. It would be revolutionary. A technology like this could provide a blueprint to build practical, hydrogen-oxygen propelled machines of all shapes and sizes. A working model built from McKane's design could be tailored to fit any commercial or residential application.

And that's not all. A greater urgency compels McKane toward his goal. He's read and re-read the constant warnings from the scientific community about the catastrophes of a warming planet. Nevertheless, the vacuum of consensus on a plan from academia, political leaders, and economic minds to rescue the environment frustrates him. Urgency and anger fuel his single-minded pursuit of innovation.

McKane knows the climb ahead is steep. Given his limited credentials in science or technology, he has no business tackling the problem of climate change. In the back of his head, a voice on loop sneers at him.

You can do nothing about it, McKane. You are presumptuous to even think so. You are not a scientist nor even a credentialed educator.

McKane tries not to run the numbers in his busy mind, to calculate the odds against him.

Maybe the PTSD has made him crazy.

Some of his family and friends think so.

* * *

His dad and mom are firmly supportive. But that's not a mystery to McKane. They are tree-huggers from way back, now living on a small tree farm next to an Indian reservation.

McKane's parents grow six species of redwoods on their land, including the deciduous redwoods that once covered the entirety of North America a million years ago. They even raise Red Dawn redwoods, once thought to be extinct. The only evidence of the tree's existence could be found in the fossil record until 1943, when botanists visiting a remote region in China unexpectedly found a valley of the huge trees. This landmark discovery led to a resurgence of the species in Europe, the United States, and his parent's property.

McKane's parents understand the power of scientific discovery, and the stakes of their son's work. They offer to stake him financially, even as McKane throws his own savings at the project.

* * *

McKane stows his equipment and carries out his work in a large garage space on loan from a kind widow. The makeshift laboratory sits on a solitary plot somewhere in a forest north of Seattle. Though he can't afford a long-term lease of such a large workshop, McKane does chores around the woman's house and acreage to make up the difference.

But rent money doesn't interest her anyway. After all, she's retired with a good pension from Alaska Airlines and owns two other garages. The shop she loans to McKane was her husband's special retreat, a massive 'mancave,' with workspace and storage for his prized RV and custom-built fishing boat. According to his surviving wife, the man often tinkered there with lights flooding late into the night. He had lived a satisfying and long life full of adventure.

In the widow's tales of her husband's life, McKane senses an echo of his own ambitious spirit.

Just after high school, the man had learned to fly bush planes across the Alaskan landscape. Decades later he took a job as a pilot for Alaska Airlines out of Seattle. There he met his wife, a chief stewardess for the airline. Years into his retirement, the man held onto his flying credentials. Though he never lost the hunger to fly, his legs eventually lost the strength to carry him up and down the airplane ramps.

After retirement, he often went fishing alone in the ocean. Ultimately, he died from health complications after a boating accident. The widow's husband had fallen into the frigid waters near Sitka and was unable to extract himself for three hours.

Though she rarely enters the garage now serving as McKane's workshop, the old woman finds comfort seeing the lights in the big shop as McKane works late into the night. It reminds her of better, more loving times. With no immediate family around her, she welcomes him into her world.

* * *

McKane's lab fills up one side of the garage. The garage offers him more than enough room for his old BMW, a computer, various small tools, and machines. He places a cache of old science books at the end of the workbench. *Light and Matter, Catastrophe Theory, Radiotron, Modern Dictionary of Electronics, Physics of the Impossible, and McGraw-Hill Concise Encyclopedia of Science and Technology* are his go-to references guiding his journey. Like the alchemists of old, McKane somehow feels there are magical secrets buried in scientific antiquity. He would not have discovered his passion to be an inventor without these old books.

It is never easy finding the very old books. Many are lost. Others are forgotten after rejection by the scientific edifice of the day. But McKane's library is built on a stroke of luck.

Rena, one of ProVault's Masters National Champions, calls him one night while he is studying in the lab. Her father, a theoretical physicist, had just died. Like McKane, Rena's father had a life-long passion for science—specifically, an interest the study of gravity. Rena's father believed very few people understood the mystery and power of gravity. And he had left behind an expansive library: a treasure trove of old books, technical notes, and manuscripts.

Rena offers the resources to McKane, who thanks her profusely.

Day and night McKane works alone in his self-styled monastery. The days turn into weeks, months, and years. He comes to believe that without a fixed daily schedule he will relapse into PTSD.

His father, the psychologist, gently echoes this concern.

* * *

He takes only coffee in the morning and then a light snack or sandwich in the afternoon. Dinner is taken out or a quick meal at his parent's house. At night, he reads. Anything having to do with hydrogen density and gravity. He devours audio books about quantum field theory and physics. He finds old, expired patents that inspire him.

McKane is ready to take the next step. The problem is, McKane lacks the academic qualifications to complete proper scientific research.

* * *

McKane has no formal training in the hard sciences. He took some biology, chemistry, and physics classes in high school. But those were more introductory surveys rather than rigorous science. And his grades were average. In art classes where he was allowed the freedom to use his imagination to create something new McKane thrived. Most of all, he excelled his peers in drafting class where the focus was converting mental impressions into precise forms, diagrams, and illustrations.

At the University of Washington McKane had elected art as a major. This path allowed him the flexibility in his class schedule to be in the track pavilion twice a day. He embraced the grueling rhythm of weight training sessions in the morning and vaulting practice in the afternoon. Every Division One scholarship athlete understands the delicate balance of managing a full-time "job" on the field while at the same time dealing with all the academic requirements.

And if your GPA drops, you get fired.

McKane enrolled in all the core art classes. He found himself gravitating to figure drawing and print making, earning a reputation as an excellent sketch artist. The drawings consist mostly of fictional creatures, battles with futuristic weapons and armors, and imaginary power reactors for spacecraft.

Given the flexibility in his academic course load for elective classes, McKane picked science courses rather than the lightweight classes others took to bump their GPAs. Physical science intrigued him more than anything else. The trouble was formal departmental approval was required to enroll in upper division science courses. This became a constant hurdle for the young art student.

McKane signed up for a 200-level Mechanical Engineering 3D Design class and passed with flying colors. He discovered the existence of another 3D Design class—a 400-level course for Mechanical Engineer majors only. He petitioned the department for a waiver to get a seat in the class, based on his interest in computer 3D coupled and his prior success in the 200-level class. The course focused on CATIA—a digital content system frequently used by professional engineers. CATIA was the industry standard, used at Boeing and throughout the aerospace field.

At first, McKane adopted the role of observer. Mainly, he listened to the lectures and avoided chatter with his classmates, having little in common with the engineering majors. For their part, they politely tolerated him.

But he overheard the sarcastic comments about the "art major" in the room. An academic paranoia soon sets in. In the back of his mind, he remembered hearing from a wise man, *only the paranoid survives*.

To his engineering classmates' surprise, McKane made the Dean's list that semester.

* * *

McKane moved on to courses in physics and astronomy, where he earned solid *A*'s. On one occasion, the *Physics of Light* professor invited him to her office. For a while, they conversed in turn about the Hubble telescope, new-found galaxies, meteors, meteoroids, and vehicles needed for future space exploration.

Before he left her office, the professor said that McKane inspired her.

8

Chapter 5: Discovery

He approaches lab work as a painter to a canvas—with an open mind patiently awaiting the illumination of creative energy. It is his biggest asset. Invention, after all, is a product of the imagination first.

McKane doesn't feel gifted. Rather, at the bottom of his psyche, he feels himself to be a unique species of human: half artist – half inventor, half man, and half ghost. Since the accident, he senses having one foot in three-dimensional reality and the other across a forbidden, invisible boundary in space and time.

He is not sure on which side of the boundary he belongs, or even which he prefers.

His soul is bifurcated in a way he can't quite explain.

* * *

The day begins inauspiciously. He is in the lab working on the same old problem: how to efficiently split water into its constituent parts. His prime project for a thousand days. He has no special plan for the day ahead. There are no insights flooding in to guide his action. He understands having no destination at all is the perfect state of mind for work like this.

He is experimenting with a new configuration of his electronic board and its interface, crafting what he thinks of as a "resonant cavity." But he is frustrated. His novel electronic scheme isn't working as planned.

McKane turns up the voltage. The amps spike up and the voltage drops. Nothing happens in the water. He can find resonant frequency with the coils and the water, but no hydrogen is produced. He tries one hundred different configurations.

Nothing works.

He pushes harder, his efforts lasting the day. Finally, he steps away from his equipment, totally exhausted.

McKane sits down and rests his head in his hands, granting himself brief reprieve from desperate impulses. Unwanted thoughts begin to intrude, preoccupying his mind and reshaping his mood. Soon, a stark realization hits him.

This is it: the end of the journey. The insights were wrong. The project has failed. It was a terrible idea. Years lost forever. There is nothing ahead.

In McKane's psyche, already fragile thoughts and feelings enter freefall.

He tries to put his mind back in a neutral gear, focusing on controlled breathing to detach from the swirling vortex of desperate thought. He waits, watching for the moment to pass. It doesn't. Instead, a no-warning flashback catapults him into a dormant memory.

It is his last panic attack years ago in the psychologist's office—the time when he couldn't move the wooden doughnut onto the right peg—trapped within the boundaries of an impossible puzzle. Now, he finds himself tangled in another enigmatic maze, with far more ominous consequences.

McKane is crying.

PTSD is in control, propelling his thoughts, reminding him *everything is pointless.*

He wonders if he should just kill himself and be done with it.

Suddenly, another vision comes to him, more powerful than the last.

Through his tears, McKane sees the ghost from the mountain. The ghost is in the lab, watching him. Once, the wise apparition stood with him at the edge of an alpine precipice. Now it has returned to be at his side on the edge a deeper abyss.

The ghost sees the machine. The ghost is in the machine. McKane hears a faint whisper.

Just do it like this.

McKane understands the ghost's command, but he cannot obey. It makes no sense. No electrical engineer would configure electronics the way he suddenly sees it in his head.

It could never work. He comes close to ignoring the ethereal admonition and giving up.

* * *

The inventor stands and wipes the remaining tear from his cheek. Disentangled from doubt, he moves in blind obedience to the soldering station to begin reconfiguring his electrical board exactly as directed.

McKane connects the new setup to the resonant cavity and energizes the system.

The device begins to pulsate with energy. The voltage is skyrocketing even though the amperage is extremely low and stable. Something strange is happening. Masses of excited hydrogen and oxygen bubbles are powerfully surging through the system.

Astonished, he wonders, *where is all this energy coming from? I've never seen this before. No one has ever seen this before.*

His eyes pass back and forth over the machine as revelation forms in his brain.

I can synthetically produce off-the-chart levels of combustible, pristine hydrogen, and oxygen gas at only 1.9 watts.

He lingers in the lab, staring at the unbelievable energy readings coming from the miniature machine. Moments later, he is at the white board doing the math. It works out. The electrical input is small—only 1.9 watts—while the energy output is powering two 5-watt light bulbs.

The weight of the discovery begins to dawn on his tired mind. The very existence on the benchtop invention changes everything.

He cannot begin to fathom the implications of the discovery.

* * *

McKane feels as though he just witnessed a birth. He wants to celebrate the event, to share the cheery moment surrounded by loved ones and colleagues.

But there is no one with him in the lab and hasn't been for five years.

A sharp pang of loneliness travels through McKane's body like a muscle spasm. It awakens a memory of another time when he should have been happy for cheating death on the mountain, but instead he was struck down by depression.

He moves to shut off the machine, then the lights. He closes and locks the door to the lab. Under the dark sky McKane traces his steps down a worn forest path, the quiet corridor to his tiny living quarters, where no one is waiting for him. With a sudden headache seemingly out of nowhere, he drops heavily onto his bed and falls asleep as if nothing happened. It is going to be another silent night.

* * *

McKane wakes early the next morning. He lingers for a moment on the edge of the bed and tries to recall every detail of the day before. A question keeps revolving in his mind: *where is all this energy coming from*? He agonizes over the steps necessary to answer the question. An enigmatic puzzle in science may have been solved.

He wonders if he can replicate the proof.

McKane returns to the lab even before morning coffee. He opens the door and steps through slowly—deliberately—like a man entering a religious cathedral. The soaring ceiling still consists of sheet rock and white paint, not large, imposing stone with ornate borders of imported tiling. Yet, it feels like he is entering the Catholic church where his sister, Dejah, attends services. He reverently approaches the newborn object on the workbench as if it were a historic relic of great significance.

He gathers up the forensic evidence still inside the machine, at first hesitating to touch the object for fear of disturbing the exact wiring configuration that produced the magical release of excited oxygen and hydrogen gas.

Would it be a sacrilege, he thinks, *if the lost Benben stone from the top of the pyramid was discovered, and I dared to touch it?*

The water is still in the cylindrical resonant cavity at center of the device. He calls it the *well*, and later he will call it something else more ominous. He sees the residual water in the well has turned a yellowish color. Before the test, it was clear well water. As if handling a priceless ancient sculpture, he carefully extracts most of the treated liquid with a pipette vial bottle and tightly secures the cap.

But now McKane notices hundreds of black particulates inhabiting the remaining liquid at the bottom of the well. He wonders, *what are these nuggets and how did they get there*? He will need to capture the remaining water and the black particles together. McKane grabs a second pipette vile and carefully evacuates the resonant chamber, planning to segregate the sediment with a soft microfilter and then set it aside to dry before further analysis. But first McKane grabs a neodymium magnet to see if the rock-like material reacts to the presence of a magnetic field. He slides the magnet around the outside of the glass vile.

The black particles start to move and then to swirl in the watery matrix, precisely following the movement of the magnet.

Then another impossible truth dawns on him.

The invention can synthetically produce magnetic rock from pure drinking water at only 1.9 watts and at room temperature.

He formulates a plan to get third party verification of the mysterious discovery.

9

Chapter 6: Following Quantum Breadcrumbs

McKane walks through the front door of Edge Analytical, a highly accredited environmental testing laboratory in Washington.

He has no appointment.

Politely, McKane tells the receptionist he wants to see the manager. She talks to someone on the conference line passing the message, then turns back to him. As they wait, McKane makes conversation with the young lady, who tells him, "It is nice to meet such a positive person. Most the people that come in here are kind of grumpy. They just want to get their stuff and get out."

A tall gentleman in a lab jacket enters the reception area, greeting McKane with a big smile. His name is Bryce. He wants to know how he can help.

"I want to have a water specimen tested. I have it here in this glass vial."

Bryce takes the vial of treated water from McKane and holds it up to the ceiling light. He says, "Oh my god, that's a lot of iron in there!" Bryce asks where it came from.

"I made it. I have a proprietary process. It is like electrolysis only 180 degrees different."

McKane realizes that what he just said will not make sense to the Bryce. Or anyone else, for that matter.

Bryce continues, "Are you putting anything into the water before energizing the system?"

"Nope. It starts with either well water or seawater. That's it."

Interested, Bryce pledges to obtain a full list of elements from McKane's sample, including the heavy particles from the golden water.

* * *

One week later, Bryce calls to report that the data values on the sample are in. McKane asks for the details.

Bryce explains that the energy process increased the iron content in the well water by 3,405 per cent when compared to the untreated water sample. He points out that this change is far more than expected from standard methods of electrolysis, even though stainless-steel electrodes are composed mostly of iron. Even more surprising, McKane's experiment only ran a few hours, not for days or weeks as often happens in other electrolysis' methods. Most startling to Bryce is that the experimental process did not require emersion of the electrodes into a dielectric spiked with KOH—potassium hydroxide—which facilitates the ionic exchange in the water. McKane's untreated source looks like unaltered local well water or pristine seawater without any expensive chemical boosters. Bryce points out that nitrogen and silica typically decrease in standard electrolysis, but these elements increased substantially in McKane's experiment.

Bryce concludes, "Whatever is going on inside your device, it's a special rearrangement process."

* * *

Next, McKane establishes email contact with Werner Kaminsky, who leads up X-Ray Crystallography at the University of Washington. They agree to meet in person at his office on campus to discuss the black, magnetic sediment left in the resonant chamber after the breakthrough discovery.

It is a day in late August when McKane arrives on the UW campus. He's 15 minutes early to his appointment with the professor. Early arrivals are McKane's habit since training in the Fire Academy. "If you are on-time, you are 15-minutes late," was the Firehouse rule.

Old habits die hard for the former firefighter.

McKane enters the Chemistry building to find the hallways empty. The fall quarter hasn't begun. For a moment, he imagines these halls heavily populated with chemistry majors engaged in muffled conversations about experimental processes like his. Today the halls are quiet. Each step on the tile floor echoes a hundred feet.

McKane's wearing dress shoes to match his dark suit and tie. He's not quite sure why he dressed up. As an artist turned inventor, McKane is new at this.

He guesses the correct path to Dr. Kaminsky's office, as there are no directional signs. At the end of the main hall, he takes a right turn, then a left. McKane discovers a number on a blond wooden door and figures he's reached the end of his search.

The door is locked. No one answers his knock. He wonders if this is the right door or even the right building for that matter. McKane waits in solitude as quiet minutes slowly pass. It seems an age until his attention is captured by the sound of a door opening nearby and then loudly slamming shut.

A middle-aged gentleman wearing casual clothes is cautiously approaching. His gaze is fixed upon McKane as if he were an intruder.

"Hello there. Are you from the government?"

McKane is confused by the salutation. *Is the professor serious? Or is it simply a joke to keep the situation light?*

With a smile, McKane glances down at his formal attire and suddenly understands the professor's first impression. He assures the professor he is not a federal emissary.

Dr. Kaminsky unlocks the door to his office. With his hand still on the doorknob, he says, "You must be McKane. Come on into my office. Would you like a cup of coffee?"

The professor offers McKane a chair. The two are now facing each other in a cramped office filled with papers, articles, beakers, textbooks, and a computer, in no particular order. As the coffee is brewing, McKane notices the walls are covered with academic awards, commemorations of scientific accomplishments and memorabilia from former students.

Dr. Kaminsky stands to open a window. It is going to be a warmer day than expected in Seattle. He turns back and asks, "Why are you wearing a black suit and tie? Are you sure you are not from the government?"

McKane realizes he missed the punchline of the doctor's earlier playful tease. He laughs. "No, I am not a secret agent. I used to be a student here. I graduated awhile back with an Art degree. Sometimes, I just like to put my best foot forward—that's all."

Something about this man suggests to McKane that the professor can be trusted, that the two of them may even one day be friends.

Dr. Kaminsky pours a hot cup of coffee for him and asks, "Now, how can I help you today?"

McKane explains that he is experimenting with a new type of electrolysis. As the professor patiently listens, McKane walks him through his process. He's dissociating water molecules into their hydrogen and oxygen parts without using chemicals or any chemical equipment. His dielectric can be well water, sea water or even ambient air. McKane describes his use of low levels of electrical input, not even enough to turn on a light-bulb. In this experimental process, he couples a centrally positioned water chamber, or "well," to a novel electronic board. The result is water molecules splitting apart profusely and at room temperature. McKane shares his hunch that the energy produced by this process could easily be harvested if he had the right equipment. Finally, he mentions the curious black sediment that is magnetic, and suggests that the sediment might be magnetite or even a synthetic meteoroid. He hopes Dr. Kaminsky can help identify the newly formed rocks.

"Did you bring the samples with you today?"

"Yes, of course." Slowly and deliberately, McKane passes Dr. Kaminsky the specimen across the table, as if he were handling a small, exotic creature still sleeping in its glass cage. Dr. Kaminsky examines the subject. "Wow, this looks crystalline."

Black magnetic rocks

McKane produces a neodymium magnet from his black computer bag and hands it to the professor. “Look, if you take the magnet to the glass bottle, the rocks will follow it.”

To the professor’s surprise, the rocks obediently follow the small magnet, racing along the inside edge of the bottle to keep up with the movement of his hand.

In the space of a few moments, Dr. Kaminsky’s mood shifts from playful to serious. He stands up and directs McKane to follow him. The professor opens the door and motions with his head toward the hallway. Leaving their full coffee mugs behind, they start down the still vacant hallway.

Dr. Kaminsky enters a large room inhabited by electronic devises of all shapes and sizes. It appears to be the Chemistry Department’s main laboratory. The former art-student-turned-inventor follows, his imagination spinning.

What I couldn’t do with all these magnificent toys!

The professor is at the lab table preparing the mysterious specimens for analysis. He first examines the sample with a light microscope.

“Look at these shapes - they are already crystalline!”

As he continues, McKane notices the excitement and urgency mingled in his voice.

“They have sharp edges. This is very cool.”

Falling silent, Dr. Kaminsky examines the object further. “It appears we are looking at a black body that deflects light, not absorbs it.”

Next, Dr. Kaminsky places the tiny rock on a sticky thimble head and into a mounting tray. He carefully loads it for X-Ray Crystallography. He explains what should be seen on the screen and why.

“The monochromatic x-ray beam hits the specimen, promoting and producing a diffraction pattern of rings. These ring formations tell us what the specimen is.”

McKane feels himself nearly trembling with anticipation. It is the feeling of a parent watching an ultrasound screen about to reveal the identity of a child. The inventor does not move.

Unconscious of it, McKane holds his breath as he watches the machine's screen.

As the ring formations begin to appear, something suddenly goes wrong. The image on the screen isn't forming properly.

Dr. Kaminsky stops the test.

"Sorry. I'll have to tinker a bit more to get an accurate reading for you. It seems the machine is having difficulty reading your specimen."

The professor pleads for a continuance of a week or more to get some more answers. McKane grants him the time. *Good things come to those who wait,* he thinks.

* * *

A week goes by. It seems longer. Dr. Kaminsky finally contacts him, sending an email message with an attached image. McKane clicks on the attachment first. The screen opens to reveal a beautiful set of rings, organized, and striated much like rings of a tree, or even the rings of Saturn. He reads Dr. Kaminsky's message.

> *It's fascinating, but I do not completely understand what the ring formations mean. This is a very complicated specimen, perhaps the first of its kind. It would take me months to determine what this is. You may want to check with the Molecular Analysis Facility (MAF) to get a better understanding. There is a nice lady named Samantha Young there that does XRD (X-Ray Powder Diffraction). She can help you identify the sediments.*

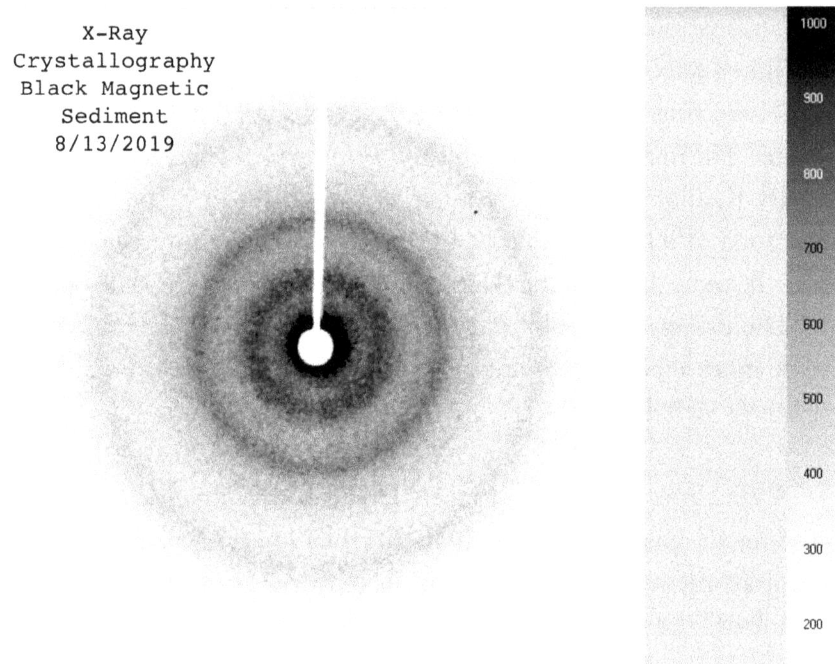

Kaminsky's X-Ray Crystallography Photo

A few days later, McKane drops off a sealed pouch of Kauai Coffee at the blond wooden door. It is, as McKane discovered in their brief conversation, Dr. Kaminsky's favorite.

McKane feels like he is entering a fold of great scientific colleagues with each step forward in his journey.

10

Chapter 7: Into Quantum Space

He has been in the lobby now for ten minutes awaiting entrance to the Molecular Analysis Facility.

The MAF, as it is known, occupies the ground floor of the Molecular Engineering and Science Institute building at the center of the University of Washington campus. It has a reputation as the largest vibration-free lab space on the West coast. He thinks the building should also receive recognition as a majestic work of modern art—a beautiful working space enveloped in engineered sheets of glass. Inside, it is big, open, and organic.

Entrance to the MAF is by a secret passage. You must be accompanied by an internal UW official wearing a security badge. The badge activates the elevator to the MAF, housed one story down. Students and visitors can take the same elevator to higher floors but not to the lower level without observing established security protocols. He realizes the MAF is not on the ground but *in* the ground.

He hasn't forgotten Dr. Kaminsky's directive to see Samantha Young after the X-Ray Crystallography failed to reveal the truth about the miniature black stones. McKane is carrying a black briefcase in his hand. Inside are specimens that may be of great consequence. He must find out.

McKane is dressed more appropriately for the occasion today, less formal, so as not to repeat the initial misunderstanding with Dr. Kaminski. While waiting, McKane notices a young lady enter the open lobby. She walks straight toward him and smiles as if they were longtime friends. In reality, they have never met in person. But McKane had sent the woman an introductory email in the second week of September.

> *Greetings Samantha - my name is McKane Lee. I'm an alumnus from the University of Washington, way back in 2007. Yikes, I know, weird that after all these years; I'm back on campus searching for answers to life... Lol. I've been trying to get some samples analyzed here at the UW. It has proven quite difficult based on the samples I have synthesized. Martin sent me to Werner, and then, Werner sent me to Tatyana. And now, after meeting Tatyana today, she is directing me to you. It's wonderful, I get to meet everyone! And I must say, everyone has been wonderful! I'd love to meet up and discuss the process and see if these four samples can be properly analyzed with the Bruker D-8 (tight focus range) in your department. I postulate that the samples will be small pits of black/brown Fe3O4 and Werner found another white substrate (not visible with the naked eye) but was unable to make out the X-Ray Diffraction scattering pattern... Werner mentioned potentially calcite... There are just so many questions because the black flakes are also strongly attracted to a magnet. I know school is about to start and "things" are getting crazy busy on campus. When would be your first available time to meet? I'm free and I care very much about getting to the bottom of this puzzling question, or at least get started... Hope this email finds you well. Thanks for your precious time. Best.*

Samantha Young, PhD, is a Materials Chemist and the Instrument Manager at the MAF. Understanding her background gives McKane additional confidence. She works with outside industrial R&D

applications and alongside the top crust academic peers solving complex measurement problems. Most importantly, Dr. Young has access to the most sophisticated electronic instruments in the world. The MAF facility is state-of-the art.

She wouldn't waste her time with McKane. There are thousands of chemists around the country she could consult with, if needed. But she believes there is something to McKane's scientific experimentation. He feels reassured that Dr. Young will work through any challenges posed by analysis of the remnant elements.

Dr. Young's office is also underground. The two of them walk together towards the elevator adjacent to the lobby. Dr. Young activates the elevator with her security badge, and they step in. Moments later the elevator opens to reveal MAF.

The entirety of it is large, appealing, modern in design, pristine, and sanitized. McKane thinks briefly that this place rivals the stories told about Area 51. He shoos away the thought, but it's a challenge. The reality of this space is more impressive than the imagination.

Dr. Young's office is just beyond the elevator door exit, a glass enclosure surrounded by many other glass enclosures. She leads him past her office to a larger room, a laboratory. Along the walls are electronic machines arranged in meticulous order. Desks, tables, chairs, and shelves have been arranged strategically so as not to interfere with the room's openness.

Dr. Young greets a colleague with a badge that reads *Scott Braswell*. She introduces Scott to McKane, mentioning that McKane is doing research on water and energy. McKane makes a quick mental note that Mr. Braswell works with electron microscopes.

After the brief introduction, they cross through the lab until they are facing two large XRD machines side by side against the wall. She selects one of them, indicating this is the machine that will analyze the specimens. It works, according to Dr. Young, by scanning back and forth over the top of the samples. Significant crystalline peaks reveal information about the composition of the subject material. She assures

McKane that this machine should reveal the identity and composition of the materials.

Her confidence gives him hope.

* * *

A week goes by before McKane receives an email message from Dr. Young. Soon after, he receives another—followed by yet another. She reports having trouble understanding the samples, especially the distilled water paired with copper electrodes. According to Dr. Young's review of the machine results, the water has iron in it when it shouldn't. And the particles are magnetic when they should be inert. Though she could make educated guesses, Dr. Young is unable to describe the identity of the black particles without more information about its crystalline structures.

She has questions for him about the origins of the material.

To McKane her words reveal a sense of bewilderment, almost frustration. And her next words seem to close the door on further inquiry. "I will return your samples to you", she says. "Anytime next week if you would like them."

McKane is crestfallen. After visits to three world class laboratories, the scientists he encounters are genuinely fascinated, but frustrated and devoid of answers. Their instruments cannot establish the identity of the sediment that appeared in a small machine sitting on a workbench in his makeshift garage-turned-laboratory in the woods.

McKane tries to comfort himself with the ancient aphorism of Confucius.

It doesn't matter how long your journey takes as long as you don't stop.

I can still do this. I must solve this, he thinks, nursing the same feeling he felt after "failing" a puzzle in a psychologist's office years ago.

Perhaps the truth of it lies farther below the surface.

* * *

McKane opens his computer, gets on-line, and goes to the MAF home page.

After rummaging through the index of names and associated specialties, he stumbles across a name he remembers: "Scott Braswell." He had almost forgotten Dr. Young's introduction of Scott on the day of McKane's first MAF visit. According to his professional profile, Scott Braswell works with the most advanced version of the Scanning Electron Microscope.

Just what McKane needs to get answers. He quickly composes an email.

> *Hello Scott Braswell. My name is McKane Lee and I'm an alumnus of the University of Washington way back in 2007. I've recently been conducting some XRD sampling data with Samantha Young. She mentioned your name and I'd like to inquire about potentially snapping a few pictures with your electron microscope.*
>
> *I've got two electrodes both measuring the same dimensions. The dimensions are 4" long and 0.50" diameter. They have developed a strange coating on them. Samantha and I have trying to determine the atomic structure. Samantha is writing up a report at this moment on her findings with XRD.*
>
> *I'd like to get high resolution visible documentation on the material on the electrodes.*
>
> *Would you be able to assist with this? I'm not interested in being trained on the equipment. I'm in the middle of research and development and don't have time to allocate to the machines. Nor would I want to accidentally screw something up on such an expensive apparatus.*
>
> *Please, let me know when you would like to touch base and meet up. I know school is about to start soon. I've got to visit the department regardless to pick up the source sample from Samantha. So, I'd like to hit two birds with one stone.*
>
> *Would you be available tomorrow? Hope you are well. And I am very much looking forward to chatting.*

Best,
McKane B. Lee
P.S. I've attached two pictures of the samples.

Having introduced his purpose, McKane waits for a response. Several days pass without a reply, and McKane wonders if there will even be a response. *Should he resend the message or make a phone call?* He senses he is on an urgent quest, but time is short, and it's difficult to be patient.

McKane braces for another disappointment.

A message from Scott Braswell finally appears in his email inbox, featuring a personal invite back to MAF. Scott reminds McKane to bring the samples, so he can get started right away.

McKane's mood soars. This time, he hopes to find a consequential truth.

Days later, McKane enters the multiplex through double glass doors engraved with the UW / MAF emblem. He turns right into the imposing lobby, where Scott is waiting to greet him and to lead the way underground. McKane is relieved to be back at MAF.

In truth, he has nowhere else to go now.

They walk together into a glass-encased room with tables, chairs, lab equipment, manuals, and the Scanning Electron Microscope (SEM). The two of them sit down side-by-side at a table facing the SEM. McKane hands Scott the dried samples of what appears to be tiny, copper-colored nuggets that he had harvested during a recent test run with well water and copper electrodes at 1.9 watts for four hours.

Scott prepares a sample and slides it into the SEM to perform the microscopy. He scans the object to select a target of interest. After locating one in the image, he increases the magnification.

"Look at the crystals! They are all facing the same way!" he gasps. "I can get even closer. This is something very special, amazing! I have never seen anything like this before!"

Turning to McKane, he asks, "What are you doing to get this?"

"I am here for you to tell me."

Copper-Oxide Quasi-Crystal Formations

After a moment of silence between them, McKane stands up, walks to a whiteboard, and begins drawing a figure. He creates a graph with a x-y axis. He inserts numbers on the graph and symbols on the board. The rudimentary drawing signifies magnitudes of mass, hydrogen densities and molecular binding energies. He stops and lays down the marking pen.

"There."

"Oh my god, you are doing transmutation." Scott says.

"Your words, not mine."

Scott continues, "This is so cool. If you have any other samples, bring them to me at once!"

McKane thinks, *Scott is sure taking to this well.*

On the long I-5 commute home, McKane notices the traffic ahead feels lighter than usual.

11

Chapter 8: Partners into the Portal

Scott Braswell and McKane partner for the next two years. McKane runs the tests in his lab, then brings the results to Scott's lab. The process never changes.

Scott uses the SEM to analyze the sediment under super high magnification and then reports back. The photos are out-of-this-world. Images never seen before. As though under superstition, the two of them never openly discuss the implications of being on the edge of a significant discovery.

They don't want to spoil the magic of discovery by saying it out loud.

* * *

Throughout their extended partnership as researchers, Scott has never seen McKane's actual tabletop device—even a photo of it. He is only interested in analyzing the molecular products of the process. Though unspoken, the two of them share an understanding that they will limit talk about the underpinning technology. Whatever it is, it already exists, sitting on a garage workbench in a forest. Their focus now is compiling and unifying a body of observations of pre-test and post-test water samples derived from the Quantum Kinetic Well ("QKW™").

McKane anguishes over proving the science of it, even though the findings are reassuring, consistent, and repeatable. In every experimental trial, the observation is that atomic elements are transmuting into an array of heavier elements. The originally sourced particles in the water seem to fall apart, or are pulled apart, while migrating to the bottom of the QKW™. At touch down, the elements have new atomic identities. Magnetite, silver, aluminum, yttrium iron, niobium, zirconium, iron, and other novel elements consistently appear as if from no discernible source.

Pen-rose octagon quasi-crystal tiling

McKane repeats the pre-test and post-test procedures time and time again. And Scott confirms the results.

Scott is staggered by another observation arising from his research. While the QKW™ effect produces many new elements, it also mitigates concentrations of certain noxious elements. The post-test samples consistently reveal markedly lower levels of deuterium, tritium, strontium, and salt. Even the electrodes are changing, sometimes growing in weight —not eroding away as in traditional Faraday electrolysis.

More breathtaking still, these transformations begin at only 1.9 watts of power. There are no artificial additions of thermal power or heavily spiked chemicals to escalate the process. Only natural well water, ocean water or distilled water are used. And the device operates perfectly at room temperature.

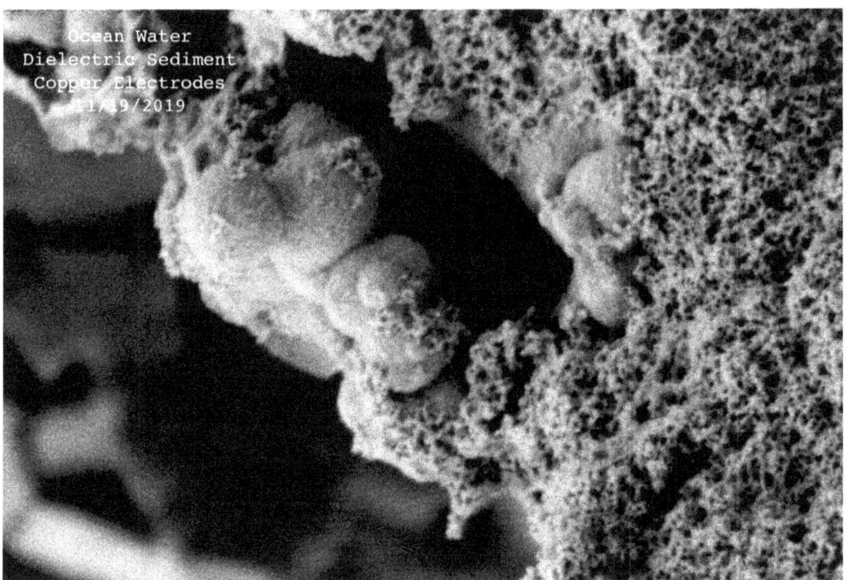

Paramagnetic nano-sphere(s) sediment

Scott can't get his well-trained scientific mind around the emerging realization that anyone who can turn on a light switch can now turn on a transmutation machine. The only difference is that it takes less energy to activate the QKW™ than to power a tiny Christmas tree light bulb.

Scott and McKane avoid another subject. They're actively testing a theory currently held in disfavor by the scientific community.

Once considered the holy grail of scientific inquiry for more than 40 years, the theory often described by scientists as "cold fusion"—that electrolysis conducted at room temperature might yield a significant energy source—was abandoned forever after decades of disappointed attempts in the laboratory. The official death certificate of the concept was issued in an editorial by *Nature* magazine a few years earlier.

Once holy ground in scientific fields thereafter became forbidden territory to any scientist wishing to preserve his standing among peers or academia.

But McKane will not accept conventional thinking on this subject. Nor can he disregard what he and Scott have now looked at under the most powerful microscope in the world—the molecular structures of previously unknown objects. Seeing new elements emerge in a quantum of space and time is nearly stupefying—a world of synthetic space rock and organic first forms. The images are both shocking and beautiful, like something lifted from a horror movie.

Or the first chapter of Genesis.

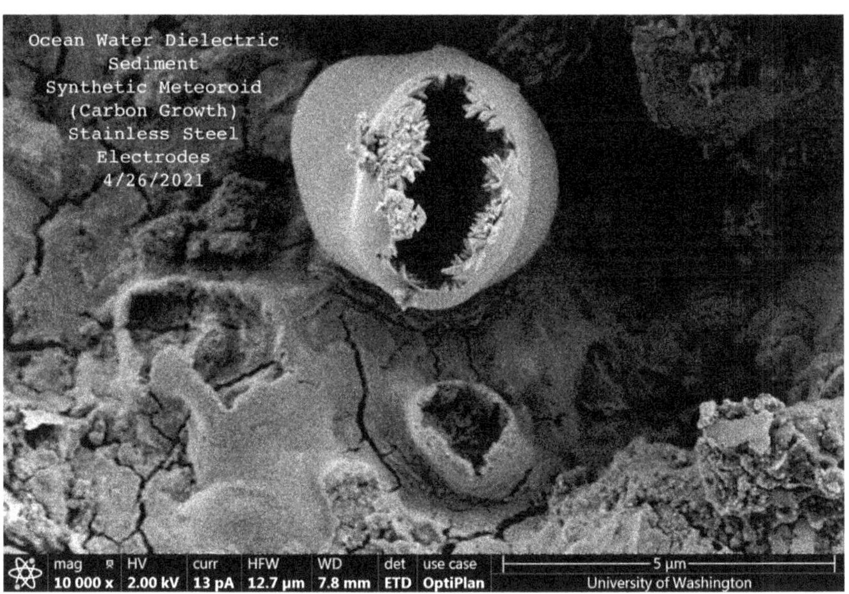

Hollow Carbon Nano-Sphere

* * *

Every few weeks, McKane had taken the journey to the MAF to deliver new samples to Scott. The drive took 55 minutes in good traffic, but up to two hours at rush hour, because of Boeing traffic out of Everett early in the afternoon.

McKane would like to meet regularly with Scott, but Covid-19 restrictions now bar his access to MAF. So, McKane begins leaving the samples for Scott in a locked box inside the lobby.

Nevertheless, the inventor and partner scientist continue communication by text or email and seldom talk on the phone. Though now confined to separate labs, they nevertheless share a sense of purpose.

Alone in his lab again, McKane feels himself to be a de facto secret agent, working on a secret mission. The problem is that no government agency, or even academia, acknowledges him or his work. Disavowal comes in many forms, from explicit rejection to the cumulative burden of economic deprivation. He hasn't had a paycheck for more than five years.

But it helps to work with a well-known and skilled scientist at the MAF, one who understands the gravity of the work they're doing.

On April 21 of 2021, Scott reassures McKane in a facetious note: "I am always available to look at more synthetic space rock from some alien tech out of the seas of Atlantis for you. I can reserve some microscope time on Monday if you like."

Hyper magnetic synthetic black rocks

12

Chapter 9: Unexpected Vistas

Contaminated water.

McKane wants to introduce the world to a future without it.

By additional research McKane comes to suspect that his invention opens the doorway to low-energy commercial water purification plants hundreds of times more powerful than existing water treatment systems. He envisions clean, cost-effective methods to remove large quantities of radioactive isotopes from spent nuclear wastewater. If McKane is right, small modular versions of the QKW™ could be installed in undeveloped geographical areas to ensure pure drinking water. Desalinization systems could even be simpler and far more proficient, using his technology. Clean energy sources would now rely on nature's own free-energy conversions systems.

And based on the strange, almost unreal findings in his lab, he believes all these breakthroughs are possible in the time span of a single lifetime.

Too much evidence is the right amount, he reasons. This is never truer than when standing in the courtroom of public opinion—and when the jury pool will consist of academicians and scientists from across the entire planet. McKane realizes that the honesty and justice of his proof will soon stand trial in the court of public scientific opinion. And he cannot shake his worry that a rigid scientific community will reach an unfavorable verdict, despite the facts on his side.

With concern about a skeptical reception of his work growing in his consciousness, he thinks of another of his father's humorous quips: *When in doubt, overreact!* So, McKane embraces his father's quip as advice. He reaches out to a national water expert before a worldwide trial.

* * *

Gerald Pollack is a larger-than-life Biochemist at the University of Washington. McKane has been impressed with Dr. Pollack's theories on the science of water ever since he read his books *The Fourth Phase of Water* and *Germs and Gels and the Engines of Life*. He is aware that Dr. Pollack was just voted as one of the top one hundred "Most Inspirational Persons in the World" by the OOOM 100. Along with most of the scientific world, he has viewed Dr. Pollack's educational videos on YouTube and TED Talks.

McKane sends an introductory email to Dr. Pollack.

> *Greetings Jerry,*
>
> *My name is McKane Lee, an alumnus from the University of Washington (2007). I'd like to meet with you sometime next week if possible.*
>
> *I've been engaged with the Molecular Analytical Facility (Nano-Technology Department) using the electron microscope and have found some amazing discoveries I'd like to share with you. I strongly believe it is right up your alley in terms of water. To be more specific, the "4th Phase of Water." Or is that the Quantum Phase of Water? I'm sure we are talking about the same thing... right?*
>
> *In the long run, if you are interested, I'd like your support of a patent pending device. It might be worth your time. Want to meet and learn more?*
>
> *Best,*

P.S. Sneak peek: picture attached.
McKane B. Lee

To his surprise, McKane's inquiry prompts a quick response from Dr. Pollack. They set a day to meet at *Zoka*, a popular coffee shop in the University District near Laurelhurst, at 10:30 a.m.

McKane arrives too early. On purpose. He wants a little time to rehearse his presentation of key SEM images of the hybrid structures. McKane scouts the venue and selects a table at the back of the café, one perfect for discussing topics of great importance. Then, he orders coffee and a blueberry muffin, which sits untouched on his plate.

There's too much excitement for him to eat anything.

Several minutes past their agreed meet time of 10:30, McKane scans the room again, looking for Dr. Pollack. He worries that his morning will end in disappointment.

At 10:50 a.m., McKane stands and makes his way to the front of the café, a weathered computer case dangling from one hand and the uneaten blueberry muffin in the other. He takes one last look around the room. His eyes suddenly latch on the famous scientist sitting alone at a table near the front door. Dr. Pollack is pensively finishing a muffin. His coffee cup is empty.

"Hello. You must be Dr. Pollack," McKane says, stepping up to the professor's table. "I recognize you from your web page at Pollack Laboratories."

"Lucky you came by. I was just about to leave."

"So sorry, doctor," McKane says as he takes a seat. He offers his untouched blueberry muffin to the professor, as if to compensate him for the loss of his valuable time.

Dr. Pollack graciously accepts. It is his favorite.

As rehearsed, McKane opens his computer to begin his 50-slide Power Point presentation. The introductory slide is a cross-section of planet earth, including a red hot, molten core.

Dr. Pollack stops him before he can advance to the next slide.

"What makes you think the core of the earth is hot? Has anyone been there? How do we know it is not ice cold? The first rule of science is to 'be careful about what you assume is generally accepted knowledge.'"

Chastened, McKane replies, "That makes sense. Magnets become demagnetized when super-heated. If the core of the earth was molten hot, then the center of the earth would be demagnetized. It is more likely the earth's core is a gigantic chunk of magnetized space rock, magnetite. You are right, we won't know until we get there."

The professor simply smiles.

McKane gets his presentation back on track, advancing to slides with selected images from the SEM analysis. He shows the professor photos of the Quantum Well, the resonant cavity where all the magic happens.

Dr. Pollack stops him again.

"Wait a minute. I know what you are doing. You are doing low voltage transmutation, aren't you?"

McKane makes no unnecessary admissions. Instead, he says, "Funny you say that. Scott Braswell over at the MAF said the same thing."

"I wish I knew how to do that. I have been trying to do that for a long time."

McKane humbly accepts the professor's comment as a great compliment.

Their conversation turns to recent UW research projects, to their unique family dynamics, and to personal catastrophes they have both experienced.

Dr. Pollack describes how the pharmaceutical industry has been opposing his research on dynamic water technologies. McKane, echoing the sentiment, explains how a group of engineers have repeatedly scoffed at his research on water and energy.

After two hours, Dr. Pollack announces he must get to another appointment. "I wish we could talk longer. It's been great. We are now officially friends."

McKane has no words, only a sense of gratitude and hope.

The professor continues, "You know, they say I am the 89th most inspirational person in the world. Today, you have inspired me."

13

Chapter 10: On The Edge of a Quantum Matrix

Soon, Dr. Pollack sends McKane a message relevant to the future of the technology and how to bring it to the world.

"You need to contact my friend Matt. He is leading a team of top scientists working on fusion and quantum computers. Mention my name to him and you will get a response," Dr. Pollack promises, hoping to fashion a path forward for his new friend.

McKane comes to learn that Matt is the Program Manager for a large research project revisiting the theory of cold fusion, the idea that experimental, modular nuclear fusion machines can be established to imitate the power of the stars, except at room temperatures. The long-term strategy is to build such devices if the technology can be established scientifically.

So far, no one has.

McKane realizes that someone like Matt might just be the type of visionary needed to propel the novel transmutation technology forward. He composes another message of introduction.

Meeting to Discuss: "A New Quantum Technology".

Greetings Matt. My name is McKane Lee. I'm an inventor and patent writer for Quantum Kinetics Corp. Our research team has been utilizing the University of Washington for testing, sampling, and evaluations of a "New Quantum Device". The reason for my letter: I recently met with Gerald Pollack at the UW over coffee last week. I showed him the device and the application(s) of the device, and he immediately recommended I speak with you ASAP. Would you be able to meet up sometime soon for a presentation of our patent pending quantum device? Please, let me know. I've also attached a little Christmas "Image" for you in this message. Enjoy! I sincerely hope you had a wonderful Christmas and have a Happy New Year! This is going to be a "Historic" year for technological discovery for humanity! Talk soon.

Best

McKane B. Lee

P.S. "Thoroughly conscious ignorance is the prelude to every real advance in science."

James Clerk Maxwell

Matt and McKane first meet online. It is in the first days of the worldwide pandemic, and the research team executives are now working from home, cautious about Covid-19.

After breaking the ice with a comment about how they might be wearing the same blue shirt, Matt gets right to the point.

"What is it you are doing?"

McKane describes his decade-long interest in water and energy. It began with designing and building a water-splitting hydrogen booster to improve his car's gas mileage. To achieve this technology, he was able to dissociate water molecules into hydrogen and oxygen using generally accepted chemical methods. But those methods seemed to be inefficient and costly. Then one day, he discovered a novel approach completely out of phase with traditional Faraday electrolysis. It was a physical approach to electrolysis based on voltage, not amperage, to dissociate the dielectric medium. In his lab, this novel process produced unreal amounts of energy at only 1.9 watts. In addition, it transmuted molecular structures into new elements. The machine has a small nuclear signature, possibly low energy, super-soft luminous x-rays.

"I call the machine a Quantum Kinetic Well," McKane finishes.

"It has everything we look for," Matt says, matter-of-factly. "The challenge is how to establish your commercial propositions."

McKane asks what he means.

Matt describes the process. As he explains, an inventor like McKane must first establish intellectual property rights. Once he has a patent utility, then he can talk about his work openly and begin to license it to industry. Different industries will use it in different ways. Early applications of the tech will be crude. The founder of the technology will need to spend time working with industry to develop uses and applications. Don't get involved with those who wish to suppress the science. Open Pandora's box. Put the science on the table. Let the marketplace evolve around the new technology.

Matt advises McKane to be sure he has others around him committed to the process—the more, the better, so it can't be shut down.

Matt further explains his belief that humans have evolved based on proof of concepts that challenge people's sense of order. The stereotyped image in the public's eye is that "experts" should be the ones offering the proof and telling the story.

"What you are doing will challenge the stereotype."

Matt continues. "Invention is hard; making a product is harder. To start, you need to make a full narrative about your journey as an inventor. You need to convince people and get them excited about a new technology. There is a value in *writing it down*. Expect people to be incredulous and dismissive of you. You won't be the first."

McKane is taken off guard by Matt's frank vision about what is about to come in the next five years on planet earth. But he feels compelled to follow Matt further down the rabbit hole.

"What you need to know is that there is a big ocean swell about to break. There is new research. Everyone who has been briefed about it knows it is coming. It will explode like a supernova. You need to ride the lightning."

Spellbound, McKane crosses with Matt into an imaginative theoretical *wonderland*.

"We need to have 100 top minds to be working on the research together. You need to decide whether you want to be part of a collaborative network or work alone."

In addition, Matt gives a warning: "Despite your significant discovery, a 'go-it alone approach' will not work. Small inventors get stuck when they are under-capitalized."

The first hour of discussion passes quickly. Matt has an urgent matter to attend to. They agree to continue the conference in 45 minutes.

* * *

McKane finds himself catching his breath. This meeting was far more consequential than even he anticipated. As he waits for Matt to return to the computer screen, McKane decides he has time for a second cup of coffee.

His imagination has been stoked by Matt's bold prognosis of imminent breakthroughs in technology. McKane lets himself wonder, *What's next?*

Matt returns to the computer conference. Their meeting was originally scheduled to end after an hour. *As a Program Manager at such a prestigious, internationally known organization,* McKane thinks, *Matt's time must be extremely valuable. Matt must have something more on his mind.*

When their meeting reconvenes, McKane notices a different tone and tenor to their conversation. The teacher-student dyad is over—it has been for an hour. Matt is now talking to him as if they were already travelers on the same path. Matt describes his past experiences with research on cold fusion and the failures of the scientific community to prove the theory. He describes the current research on quantum computers, and technical challenges encountered. Matt wonders out loud if the quantum computer will be able to teach the world how to build cold fusion machines.

McKane offers him a riddle. "Have you ever thought that the 'cold fusion' device *is* the 'quantum computer' you are looking for?"

Matt hesitates a moment before saying, "I never thought of that."

"Remember, Matt, the shortest distance between two points, A and B, is not a straight line, it is a 'singularity.'"

***

After the meeting, McKane concludes Matt is a modern visionary, and as history proves, great prophets are often unappreciated in their own time.

McKane's machine can do many wonderful things. The inventor is convinced of that more than ever. And now he has a blueprint to bring his invention to the world. Matt has handed him the key, by way of powerful advice, to unlock many doors on the way forward.

As always, McKane walks through the next portal, not knowing if peril or promise lies beyond.

McKane wonders how it will feel to live in Wonderland.

14

Chapter 11: Maxwell's Demon

McKane reaches out to a national laboratory in Colorado to ask for assistance in analyzing the energy capacities of his invention.

This time, in keeping with Matt's advice, he makes a full disclosure on the front end of his request. He's testing a new form of electrolysis, one that produces abundant amounts of hydrogen and oxygen fuel at only 1.9 watts. The process relies on voltage, not amperage, to dissociate the water molecules. After a month of promising discussions with scientists and administrators, who initially seem quite enthusiastic about the research, the stream of calls and emails from the national laboratory suddenly dries up, without explanation.

Undeterred, McKane reaches out to a large laboratory in Texas, making the same pitch. "We don't know who you are," the lab's representative says bluntly.

Thereafter, no one at the Texas laboratory will return his calls or emails.

* * *

Facing several slamming doors to his appeals for laboratory help, McKane begins discussions with a scientific team at a university in British Columbia, the same university where he attended two years of graduate school. They express a measure of interest in the project.

However, they express concern about not having the current "bandwidth" to undertake such a study. Plus, with Covid-19 restrictions still in place at the Canadian American border, it is not possible for him to establish a working relationship in Canada.

McKane makes a presentation to a large group of engineers in Seattle, setting forth his theories and diagrams of the quantum machine. He lays out his bold vision regarding future applications of a new technology.

Before he even finishes, the professionals are making sarcastic comments to the artist turned inventor. The gallery makes catcalls, and someone even shouts-out "bullshit."

But McKane knew this was coming. He had been duly warned, by both Matt and Dr. Pollack. Still, the defamatory comments ring hot in his ears.

McKane reaches out to a good friend and faithful ally, an immensely intelligent woman. He asks for her counsel at an inflexion point in his life, reminding her he values her opinion above all others.

"Should I keep going?" he asks.

When she responds, he can't believe her words.

"You should have never started this project. It is impossible. It's been too painful for you. You have given up everything. You should think about going back to school now and start over before it is too late."

Her vote of no confidence lands hard on his ego. Repeated disappointments drag on McKane's confidence.

He fears he's slipping back into a loop of obscurity and defeat.

* * *

McKane strikes up a dialogue via email with an astrophysicist at a prominent university, hoping to engage the esteemed scientist in a discussion about solar fusion, the formation of cold bodies in space, and the chemical composition of meteoroids. When McKane offers a cursory overview of his research and the mysterious discovery in his garage-turned-laboratory, the professor initially seems interested. But when McKane turns to discuss a potential new form of plasma energy, the physicist quickly dismisses the matter as "nonsense"—and McKane's work along with it.

McKane consults with another contact over lunch at *Anthony's* in Edmonds. The man works as a scientist at a nuclear facility in eastern Washington. McKane gives an abbreviated version of the speech he made to the group of Seattle engineers.

Inadvertently, he uses the word "Bremsstrahlung," a German word for electron braking radiation.

The nuclear engineer says, "What is "Bremsstrahlung?"

"Don't you know?"

The nuclear engineer gets out his phone and looks it up. "Oh, yes, now I remember reading about that in a textbook."

The engineer looks back at McKane.

"If what you are telling me is true, and you figured it out, then I am sitting across the table from the richest man in the entire world. What is your secret?"

McKane replies with a story.

"Do you remember the famous thought experiment by James Clerk Maxwell? It is a parable about an opening doorway between two compartments full of gas particles. There is a tiny, imaginary creature who sits at the door. Every time the creature sees a fast-moving particle approaching from the left-hand side, it opens the door and lets it into the right-hand compartment. And every time a slow-moving particle approaches from the right, the creature lets it into the left-hand compartment. Pretty soon, the left-hand compartment is full of slow, cold particles, and the right-hand compartment grows hot. So, what Maxwell did was create a puzzle within a paradox, a mechanism that defies one of the laws of the thermodynamics, entropy."

"Okay, what's the point?" asks the engineer.

"What if the imaginary creature controlling the portal is a basic electronic switch?"

"Bullshit."

McKane picks up the bill and pays the gratuity on his way out of the restaurant.

15

Chapter 12: Overcoming Barriers

It takes him several days to comprehend what the incredulity of a seasoned nuclear scientist, an astrophysicist, the enclave of Seattle engineers, and two regional laboratories all add up to.

McKane is an unknown inventor and an uncredentialed scientist, and as such he will never hold their respect. McKane may have made a marvelous discovery, but his novel technology will never be accepted by the scientific establishment if it is *his* voice alone advancing a new form of energy.

* * *

McKane knows what it's like to be shunned by your superiors, even if you have remarkable skills of consequence.

He remembers his emergence onto the pole-vaulting scene when he was an unknown high school freshman competing at the District Championship Meet.

His competitors in the finals were juniors and senior students with established athletic backgrounds, school records, and names affiliated with years of feted accomplishments in track and field. All were competing for invitation to the State Championship scheduled in two weeks. Only the top three would qualify.

He recalls the glory of that day—and the pain. He remembers waking too early on that May morning, sparkling sunshine flooding his room from around the edges of an opaque window shade. That sun told him all he needed to know—the afternoon air would be warm in the Stanwood stadium. There would be little if any headwind from the southeast, and even if the wind picked up from the northwest, it would only propel him faster down the runway, leading to higher heights.

In his memory, McKane sits on the edge of his bed, unwilling to move further until visualizing once more his well-practiced technique on the runway and the strange gymnastic moves he will make in midair. He replays Sergei Bubka's powerful takeoff in his mind's eye.

Already, he is feeling slightly nauseous. He won't touch the gourmet breakfast kindly prepared by his father.

Before McKane leaves to catch the school bus at the end of the long country driveway, his father reminds him there is a chance that the UPS might deliver the new, stronger UCS Spirit pole today, sometime before the meet. McKane had spent all his savings to purchase the pole. His dad will bring it to the stadium if it arrives in time for McKane to use it.

At school, the freshman boy is distracted. He cannot concentrate in first period French class, despite Ms. Weingarten's daily effervescence. He only passively engages with his new project during Art class. He may be only fifteen years old, but he has other, more momentous things on his mind. Inexperienced and naïve, he dreams of local immortality.

All the Arlington track athletes with qualifying marks will be automatically excused from afternoon classes. The team bus is scheduled to leave campus at one o'clock. McKane elects the early lunch period in the school cafeteria. There is no time for socializing or happy chatter. He has little appetite, still consumed by nervous energy.

Before boarding the bus, he feels a strong compulsion to empty himself of the sparse remnants of lunch. He does not do so, the threat of public humiliation too great.

There are hundreds of people inhabiting the Stanwood stadium. It is filled to the rafters. Some are his friends, others family. Mostly, these onlookers are strangers who have never heard of the newcomer. His father is there, along with his older sister, Amorah—a track star in her own day at Arlington High School. McKane's mother is not present. She is working her regular schedule at the hospital in the L&D department, where she cannot be excused during a scheduled delivery.

The pole vault runway is positioned squarely in front of the stadium seating. Twelve vaulters have already been eliminated from the competition, missing at various heights on three consecutive attempts. The crowd is boisterous and sit on the edge of their seats, having already been thrilled at amazing vaults by veteran vaulters—and the one novice.

Along with McKane, only two other vaulters remain in the competition. One is his good buddy from Stanwood High School. The other is the event favorite, a muscular star quarterback on the football team and former two-time winner of the district vault competition.

McKane slips off his gray and blue warm-ups and stands at the end of the runway. He is falling behind in the competition, having more misses at lower heights than his two remaining rivals.

His name is announced over the stadium speakers, giving him just two minutes to complete his last attempt. The bar is set at what would be McKane's new personal record. In his hands is the new, stronger Spirit pole he has never used, the one dropped off by his father. The UPS driver had arrived with the package just in time before the meet.

McKane is taking a huge risk, not knowing if he can handle the longer pole without a single practice attempt.

His focus is the crossbar, but McKane is also keenly aware of his audience. He turns to face the crowd, waving his right arm over his head, pleading the crowd for more energy. He senses they are with him, invested in an underdog's last chance at victory.

He is an athlete but enjoys the theatre of it all.

His run is powerful and purposeful. He presses the pole above his shoulders and leans slightly forward into its strength, initiating a powerful pole plant. The precise and infinitesimally quick movements of his body in the air leave the crowd in a frenzy as he flies two feet over the crossbar.

Moments before clearing the bar...

His gutsiest move ever has paid off. Only his father and his co-competitor from the Stanwood squad understand what has just happened and what risks he took to triumph in this way.

The remaining vaulters fail at the height, elevating McKane to champion.

But that is not enough glory for the day. McKane decides to up the ante. He races over to the officials and asks them to set the bar even higher. Another foot higher.

The once clamorous stadium falls eerily quiet, as if they are witnessing something of historic significance. A young talent, previously unknown to them, is bursting onto an athletic field for the first time.

Once again, McKane plays to the crowd, clapping his hands above his head, inviting their combined energy to assist on his last attempt. The bar is set well beyond his prior record. When he clears the bar with ease in picture-perfect form, the crowd again erupts. He turns to acknowledge them with two raised arms, generously sharing his victory with everyone present.

Out of youthful enthusiasm or inexperience, he gives a final salute and then performs a celebratory back flip off the back of the vault pit, a type of gymnastics seldom seen in such an arena. The result is a crescendo of approval from the stands.

With congratulatory athletes surrounding him, lifting him high on their shoulders, the Arlington head coach and the team's senior captain push through the throng to confront McKane. Their unfriendly faces are matched with a hostile message. Openly, the coach and team captain confront him, saying that he may be an amazing athlete, but he has embarrassed the entire City of Arlington with his showmanship and arrogant stunt off the back of the mat.

Confused and embarrassed by the rebuke, McKane gathers his equipment and quietly leaves the stadium. He returns home in the family Suburban, instead of taking the team bus back to the campus.

* * *

A month later, in a crowded high school auditorium, the Arlington track team is introduced. They receive ecstatic praise from the entire student body. The team's many accomplishments and records are reported, including their impressive performances at the State Championship in Tacoma.

But no mention is made of the notorious young vaulter or of his legendary jumps in the Stanwood stadium.

16

Chapter 13: Into the Nation's Laboratory

Pained by a sense of rejection from the mainstream scientific community, McKane determines that he must tune out all voices opposing the new technology. To move forward with any success, he must surround himself with open-minded, supportive people.

He chats with Scott Braswell at the MAF about his frustrations in getting third party verification of his machine. Scott suggests reaching out to Pacific Northwest National Laboratory ("PNNL"), a national scientific facility and subsidiary of the Department of Energy. Scott says the PNNL campus is close by in Sequim, only a ferry ride and modest drive away from Seattle. To Scott's knowledge, the University of Washington and PNNL have a long and established relationship. Many of the scientists at PNNL either attended the UW as students or worked at the university as professors.

Scott provides an unlisted PNNL phone number to call, promising he will vouch for him whenever needed.

The same day, McKane calls the number. No one answers. An anonymous voice invites him to leave a message.

Hello. My name is McKane Lee. Scott Braswell over at the UW gave me this number. I have been doing some research with electrolysis that is being studied at the MAF. I need to get some numbers and feedback about the process. I live close by. Please let me know.

Two weeks pass with no answer.

McKane is convinced there won't be one.

While brainstorming an alternate source of verification, however, he receives a call from a young lady with a European accent. She indicates that she is calling from the national laboratory. For two weeks, she has been trying to place his project with the most appropriate agency.

"You need to contact the Oceanography Department. They will get you going. They know you will be reaching out to them directly."

The woman shares the number for the scientist at the national oceanographic laboratory chosen to lead McKane's research project. The facility is located on the shores of the Strait of Juan de Fuca in the Pacific Northwest.

* * *

They first meet by video conference. The lead on the project appears to be standing on the deck of a fishing boat in the middle of the Pacific Ocean on a beautiful summer day. It is an inviting scene to McKane, who has fished for salmon off the Washington coast on his father's boat since he was a boy.

"So, I assume you like fishing?"

"Yeah, especially for Coho salmon out here in the Strait," says the scientist.

"Me too."

"So, what is your project?"

McKane explains that he is working with a new type of electrolysis. He's trying to nail down what is happening inside the device. He references the energy production of the technology, and the curious sediment the electrolysis leaves behind. McKane wants to know about changes in the isotopes of the water. He suspects the synthetic device can efficiently alter atomic isotopes with little energy input.

But he wants the details laid out in a scientific publication, if possible.

"Can you do that?"

"Oh yes, our lab can definitely do that, no problem!"

The next time they meet, McKane finds himself standing in the John Wayne marina, near the entrance to the research enclave. Because of Covid-19 restrictions, non-authorized persons are not permitted into the sprawling federal facilities. The facility is fringed with tall fences, barbed wire, and ominous warning signs forbidding entry.

The team leader and McKane work out an approach to the project, agreeing to work as strategic partners to compile and unify observations of pre-test and post-test water samples derived from McKane's Quantum Kinetic Well.

They set up a paradigm for a series of experiments. The Leader obtains water samples for each test, using two sources for the dielectric: well water from an outlying area and seawater from the Strait of Juan de Fuca, the purest in the world. McKane meets the project Leader up at the Marina parking lot every month for a year. The scientist delivers large containers of fresh and salt water in sanitized carboys along with pristine equipment to collect the treated samples.

McKane then runs the tests in his lab and returns the treated water samples.

At that point, lab personnel, all with advanced degrees in oceanography or engineering, use the most sophisticated instruments on planet earth to study both the treated and untreated samples. The Lead marshals the evidence and collates the data into summary form. As a true experimentalist, he simply describes the tests and reports the results, without speculating on the underlying chemistry, biology or electronics producing the results. This way the findings should be unassailable.

The body of observations compiled by the national laboratory prove to be exceptional.

On first impression, the data overwhelms McKane. Complex and highly extensive, the data will require more than a single journal article to relate.

It may require a lifetime of journal articles.

He decides to limit the first publications to the exotic findings most likely to redress the most urgent environmental challenges: clean energy, universal pure water, and carbon sequestration from the atmosphere.

17

Chapter 14: Fusion Unto the World

Inspired by the first round of scientific tests, tests which confirm the invention's capabilities to manipulate atomic isotopes at a low energy, McKane takes his invention to the United States Patent and Trademark Office ("USPTO") and Canadian Intellectual Property Office ("CIPO") at the same time.

He hopes to put a timestamp on the science of his machine in North America as a starting point. After writing and filing a lengthy "White Paper" describing the concepts implicit in the new technology, McKane submits official applications for utility patents. In the patents, McKane discloses the invention's design and function, even providing diagrams on how to build the machine. He lists thirteen formal claims about its operational capacities and forecasts the far-reaching applications of the device.

McKane sits and waits. Time seems an enemy, interminably postponing fulfillment of a dream. He realizes it may take two to three years to complete the patent review process, or to even get a visit from a patent officer investigating the invention. Many patent applications are denied. Other inventors go to Federal Claims Court to plead the validity of their inventions. Some utility devices are inexplicably sequestered, like Hollywood's version of the "Ark of the Covenant" in *Raiders of the Lost Ark*, buried forever in some secret vault for national security reasons.

Five months go by. It becomes difficult for McKane to keep hope alive, given the odds against him.

While on a brief visit to Kauai, he receives an unexpected phone call from Ohio. It is his patent lawyer, leaving a voicemail.

> *This is Roger Emerson. I'm calling at 1:30 on Friday the 24th, that's 1:30 Eastern Time. I just want to tell you just got word about five minutes ago from the U.S Patent and Trademark office that they are approving the patent application we filed on McKane's Quantum Kinetic Well. So, that is exceptionally good news and normally we would have another six-or-nine-month fight on this, but they approved it right away, which I think is a testimony to how different McKane's ideas are than what's out there. So, I want you to know that.*
>
> *I know you recently sent me an email asking the status on things. Before I could even check on the status, this came in. So, your timing was great. We just got word a half hour ago from the USPTO. I just got the word only five minutes ago. So, congratulations. We are sending out our normal reporting letter along with the form letter that gives you all the details. But I just wanted to let you know right away.*

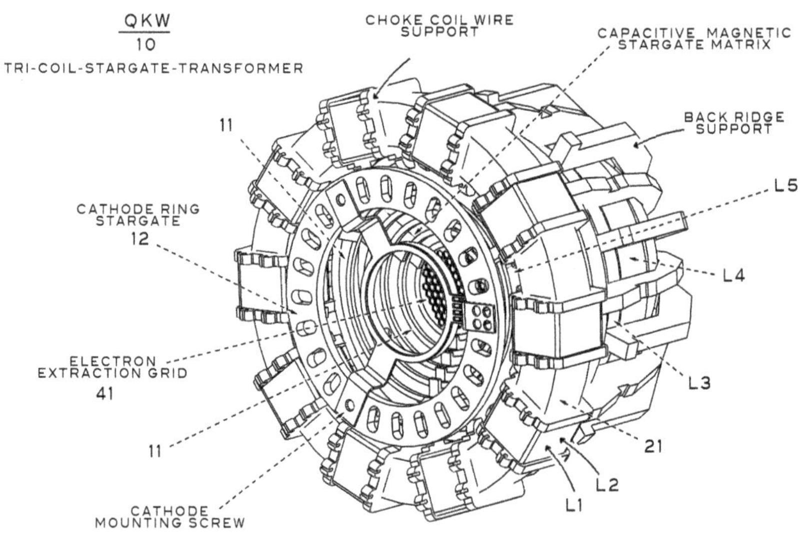

The patented Quantum Kinetic Well (Arc Reactor)

McKane cuts his vacation short and sets up a phone call with his lawyer to garner more details. As his lawyer relates, the patent officer issued the patent quickly because it was clear McKane's device "worked" and that there existed "no other devices" the same-as, or substantially similar-to, the subject invention. The patent officer expressed personal consternation that scientists had not yet discovered this technology.

McKane receives the official paperwork approving his patent. It is overwhelming. He asks his patent lawyer if there is any further verbal communication with the patent officer. There is. It comes in the form of a personal request, or perhaps even a prayer.

Go teach this to the world.

He contemplates the significance of the unofficial message. He reads the official patent award letter over and over, absorbing its significance. *Could it be that a synthetic marvel that opens a domain of boundless clean energy has appeared just in time to save humanity's home planet?* The magnitude of the commission, and its urgency, suddenly strikes home with McKane.

There is no time, or need, to wait for decades of further scientific research and analysis. The great world powers are facing off against each other to see who will capture the last drop of ancient oil. Emergency use authorization is required to stop the madness.

He reasons it can happen with a new energy source at hand.

McKane calls Roger Emerson, his lawyer, again. He wants to file for additional patents in Great Britain, France, Australia, New Zealand, Canada, and the United States, as a start.

McKane catches Roger up on a recent success in his own lab with another invention that has been in the works for a year. The patent narrative for this device is already drafted and ready to go. He explains that the new device can be coupled to the QKW™ to greatly increase its power. As an injector system, it can be applied as a retrofit technology for use in internal combustion engines, commercial airline jets, heating furnaces and rocket propulsion thrusters. The injector system utilizes hydrogen, the perfect renewable fuel for all industries because there is a natural liquid storehouse of hydrogen–water.

McKane points out that the combined thermal explosive force of hydrogen and oxygen is astronomical. That is why most existing rocket thrusters rely on cryogenic oxygen (LOX) and cryogenic hydrogen (LH2) (solid fuel) as the fuel source within a combustion chamber. But McKane's injector fuel system is sourced differently—it's not reliant on cryogenic oxygen and cryogenic hydrogen. Instead, the device harnesses the ferocious atomic-power yield from room temperature water (water-mist as fuel), vectored for rockets thrusters or any internal combustion engine on demand.

Roger agrees to start work on the injector patent. As he does, the Super Lawyer receives an important message. It is from the Canadian Intellectual Property Office. McKane's application for patent of the Quantum Kinetic Fusor, the fusion device itself, is coming through. Approval of the invention comes with a familiar message from the examiner: *The invention works. There is no competing technology. All claims are allowed.*

Roger updates McKane on the good news. The inventor, and his lawyer, are again startled by the swiftness of the award. Precedent in patent approval suggests it could have taken years to gain approval, if ever. McKane's expectation of court battles and governmental sequestrations dissipate into thin air.

Examining the patent paperwork, the inventor notes a striking similarity between the Canadian patent officer's message on the QK Fuser to the United States patent officer's message on the QK Well.

Different countries . . . same message.

With a flush of humility, McKane considers the prodigious implications of this second patent. *It is now possible to couple the patented QK Fusor with the patented QK Well, together fortifying the gateway to nature's abundant energy domain.*

On the heels of the CIPO revelation, McKane receives another game-changing message from his colleague at the national oceanographic laboratory.

Their most recent data suggests that the Quantum Kinetic Well may perform sequestration of CO_2, removal of CO_3, pH enhancements, water purification, and biological stimulation in the form of protein-matter growth, humic material, and phytoplankton fluorescence. Furthermore, the device may reduce radioactive elements such as tritium, strontium, and other noxious elements in the seawater. It even appears the device may stimulate a full carbon-cycle within the treated dielectric, something no one thought possible.

* * *

With a whirlwind of momentum behind him, McKane applies for the X-PRIZE Carbon Removal project sponsored by the Musk Foundation. This application marks McKane's first opportunity to *teach the tech* outside the confines of his limited social media platform. McKane understands there is little, if any, chance his application will win, or even be acknowledged by the team of reviewers. There are a thousand other applicants. Most are better funded, politically connected, and far more professionally resourced.

After all, McKane is not even a trained scientific writer. He speaks with the voice of an artist, not a nuclear engineer. What foundation board would even pay attention to his propositions?

Still, McKane has secured every previous victory on his journey only through stubborn, patient persistence. He will no longer refuse to explore every avenue available to him, or let disappointment deter his ambitions. McKane determines to use the X-PRIZE application process as an opportunity to spread the word—one company, one patent, one person at a time.

His submission is intriguing and voluminous. Backed up by research from a national laboratory and a prestigious university, McKane assesses that the goal of carbon removal is possible by relying on the resonant quantum transformer to recalibrate nature's own carbon-cycle. By strategically placing commercial sized units of the Quantum Kinetic Fusor and the Quantum Kinetic Well where the largest rivers meet the oceans, biological stimulation would progressively normalize marine CO_2 levels through a recycling process.

The approach would function as an atmospheric time machine, in effect taking the environment back in time to when it was pure, clean, and supportive of all life. The synthetic machine would also chemically increase the ocean's capacity to absorb carbon from the atmosphere and safely store it within its depths, where the life cycle would begin again, a world reborn.

18

Chapter 15: Hollywood Calling

Before McKane became the inventor, he was an avid consumer of science fiction literature and film. His favorite fictional character was *Iron Man*, Tony Stark, first introduced to the world by Marvel in 1963. Comic book lore depicted Tony's father as a notable inventor who, inspired by the idea and hope of sustainable fusion power, developed an electro-magnetic machine that could provide limitless clean renewable energy to the world. Although the project ultimately failed to deliver the promise of fusion power on a large scale, its underlying technology was used by Tony Stark to build a modular version of a fusion power plant–the Arc Reactor, the source of Iron Man's secret powers.

Over 50 years of superhero comic book tales and blockbuster movies, the saga of Iron Man's Arc Reactor has inspired many real-life inventors to follow the same dream of fusion power for the betterment of mankind. Recently, MIT has been working on a reactor that would function like the arc reactor, although the science team avoids using the same name.

* * *

McKane can't advance the adoption of his working version of the Arc Reactor without a comprehensive plan to *teach the tech.*

As a child, he learned in Sunday School that there was an original Great Commission, time stamped 33 A.D. This brief dictate of the Christian messiah embodied a resurgence of hope after an apparent tragic ending to a promising story of universal redemption.

Two thousand years later, the world seems locked in a state of perpetual conflict over dwindling energy resources. Perplexed by the dangerous state of international politics, McKane wonders if another critical commission is needed. The first Great Commission was propelled by word of mouth across ancient empires to the ends of the earth.

The second must spread more quickly, with a sensitivity to the impending environmental crisis facing humanity.

McKane resolves to work with anyone willing to accept the data without muffling inquiry based on their first impression, or the biases of prevailing scientific dogma. Advancing the technology will involve a cast of supporting characters, a massive funding effort, wonderful machines previously dreamed of only in science fiction, and episodes of joyous triumph along the way.

But McKane has no idea what this yet-to-be-assembled team of scientists, spokespeople, and investors will look like. To temper his expectations of positive reception from audiences across the country, McKane continually reminds himself of the implausible, *avant-garde* nature of the work he's representing.

Nevertheless, he hopes for an infusion of new energy and perspective into his work from a supporting cast of enthusiastic partners.

* * *

One day, he is meditating after a long day in the laboratory. It is a practice learned in the Happy chapter of his life—before the accident.

McKane thinks of meditation now as a non-traditional therapeutic practice inoculating him from long term PTSD. It provides him stress reduction and timeless moments of mindfulness, but rarely offers him fresh insight or revelations.

But today is different. While meditating, McKane sees an image of a face and the sound of a name within a whisper. The vision is of his old friend Greg from ProVault NW.

McKane knows that he must reach out to Greg.

* * *

McKane messages Greg in the morning. They reminisce by phone about their good times at the vault club. Eventually, the conversation turns to personal updates and current events in their lives.

McKane asks about his friend's situation after college. Greg reports working with Russell Bobbitt, the actor and legendary prop master for Marvel and Disney. Russell's movies include *Guardians of Galaxy, The Avengers, Iron Man, Thor*, and many others futuristic thrillers. Russell designs and builds original movie props by merging fantasy with reality. His products capture the public's imagination about science, fictional heroes, battles with exotic weapons and armors, and imaginary spacecraft powered by fusion energy.

The Arc Reactor used in the *Iron Man* movies is one of his fantastic creations.

Stunned by coincidence, McKane describes his recently patented invention that generates a magnetic field while ionizing the ambient air at 12v input, like the theory behind the fictional Arc Reactor.

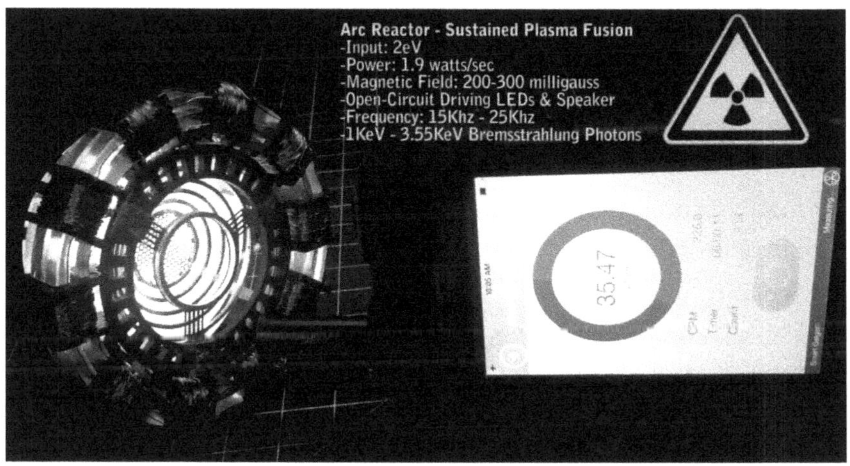

The Arc Reactor sustaining plasma fusion (Bremsstrahlung Radiation)

Excited, Greg promises to inform Russell of McKane's work.

It seems days, but it is only hours, until Greg reports back. Russell wants to talk to McKane and soon.

For a moment, McKane feels some concern, a twinge of paranoia. He considers the risks of a meeting. *Will these massive companies sense I am infringing on their intellectual property rights? Is Marvel about to marshal an immediate "cease and desist" court order on the new technology? Do I need to call my patent lawyer?*

Sensing a need for caution, McKane asks for more context from Greg before agreeing to a meeting. He inquiries about the reason behind Russell's request to meet, confident Greg will tell him straight.

"When Russell learned about your patent, he seemed staggered," Greg says. "He just wants to talk with you to find out how it works."

McKane considers the potential advantages. *What better way to communicate to the world than through large international corporations capable of the mass production of books, TV shows, press conferences and blockbuster movies?*

"Let Russell know I would be delighted to meet with him at his convenience," McKane says, taking a first confident step into an entrepreneurial world of communication by film and storytelling.

Russell and McKane meet by Zoom call. Russell's warmth and wit immediately reassures McKane that this will be a friendly meeting, focused on the new technology and its applications rather than any concern over intellectual property rights. Russell is not interested in such IP matters, in part because no one has patent rights on a functioning Arc Reactor.

No one, of course, except McKane Lee.

Russell explains the reason for the meeting.

"I see what you have done with your technology and the Arc Reactor – very smart. I love your story. I was excited to learn someone finally figured out how to make a real Arc Reactor. I have been waiting for years! I knew it was possible. When I heard about your patent, I called my father-in-law, Michael. He is a recently retired executive with the USPTO. And a member of the Intellectual Property Hall of Fame at the federal agency. Michael was surprised to hear news about your invention. He revealed few inventors are awarded a patent in just five months, especially for a new kind of technology. He concluded "somebody must be looking out for these guys."

"Interesting. What can you tell me about your involvement with the Iron Man story and the Arc Reactor?" McKane asks.

"The movie version Arc Reactor was designed and built in my shop after talking with scientific consultants and brainstorming with Robert Downey, Jr. The ideas behind the reactor were electron capture, electromagnets, and fusion."

"Some very special electromagnets, I assume."

"Exactly. The scientific premise is controlled fusion energy as the source of Iron Man's secret power. It is also the only way to sustain Tony Stark's life after he incurred a serious injury in warfare. The device controls his every breath and protects him from embedded shrapnel, lodged too near his heart."

"I remember the part about the device pulling the shrapnel away from Tony Stark's heart. I will have to work on that," McKane says, smiling.

Russell further explains his work. "It all starts with a scientific proposition, even if it is never developed into real technology. The public bought the idea of the Arc Reactor, literally. The toy version was mass-produced for sale. Disney's target audience has always been the 7-year-old child. Children believe in the possible. Every child who buys a toy Arc Reactor is affirming the possibility of a future that includes it."

Russell turns the conversation to McKane's journey.

"I hear you locked yourself up in a garage for five years and worked on your project."

"That's all true."

"I am so envious. I wish I could do that. I am an inventor, too. I dream of escaping into the solitude of my laboratory for five years and design an all-electric car that travels above the ground. Maybe I'll do that when I retire, not too long from now," Russell says.

Russell asks about specifics. "What about the energy output of your machine? Is it capable of operating a large manufacturing factory or a small electrical power plant?"

"In the laboratory we work with a modular, tabletop unit. To run a small power plant, you would increase the size of the electrodes and wiring configurations to extract more electrons from the water, thus creating more electrical current. The extent of energy harvesting is only limited to the size of the electrodes. This means our quantum system can also be engineered to run large power plants."

"How much does it cost to convert over to your electric system?"

"It depends on the existing business infrastructure. Many are ready-made for it now; so, it's plug-and-play. Some will require more redesign work. Ours is a retrofit technology for use in any energy demand situation, including internal combustion engines, rocket propulsion thrusters and regional power grids."

"Could it be used as a car battery recharging station?"

"With our prototype we generate amazing amounts of energy at only 1.9 watts. Other versions could be designed to power batteries of all shapes and sizes."

Russell cannot hide his enthusiasm for McKane's work.

"What you are doing is so much bigger than science fiction and selling toys. It is even bigger than a working version of the Arc Reactor. It is a system to power the world, a new form of electricity. This will be scary to some, exciting to others. Now you need to market your invention.

Think about how Elon Musk became successful. I was in the room with him when he first said his rocket was flying to Mars in 10 days. Of course, that came true. Elon explained that when he was a young man, he pledged to never set off a rocket smaller than the last. He has been true to his promise. Elon has promoted his business through smart applications of cutting-edge science coupled with the power of imagination. You are in a similar position, except for the whole gazillionaire part. Your invention can change the plane of history. It seems you are the z-axis of electricity."

McKane feels as though he has landed a crucial and understanding supporter of his work in Russell Bobbitt.

Russell affirms McKane's impression. "If you want our help, we are all about that."

19

Chapter 16: On the Z-Axis

After his Zoom call with Russell, McKane is propelled into fresh reflections about his work. He understands Russell to be making bold predictions about an emerging technology.

McKane lets himself wonder.

Should this device really be considered a new form of electricity? Does anyone really merit being called 'the Z-Axis of Electricity'—let alone a newcomer to the world of scientific discovery?

McKane understands, along with all first-year geometry students, that Cartesian Coordinates in three-dimensions allow for ordered triples of real numbers to be displayed within three perpendicular planes intersecting at one point, the origin (0,0,0). This three-dimensional space, created by merging a z-axis into a simple x-y dimensional plane, allows for a vastly more complex and varied visualization of possible locations.

Russell's reference to the "z-axis of electricity" staggers him with a newly considered possibility.

Is it possible the Quantum Kinetic Well operates at a location along an infinite axis in space never identified, where energy access is as elementary as engaging a resonant electronic switch, just as predicted by Maxwell's thought experiment?

Stunning realization marks another inflection point on his journey. McKane senses the practical effects of his next decisions will surely be momentous. He is sure about only one thing: the vast capabilities of the tiny machine existing on a table in his laboratory.

Under the influence of newfound support from a voice in Hollywood, McKane is bursting at the seams to take the technology into development and marketing. Still, he does not want to make the mistake of the novice traveler in an unfamiliar wilderness, a hasty decision leading to sudden catastrophe.

So, he turns to someone more familiar with the landscape of technology development and commercialization, messaging his friend Matt. Luckily, the entrepreneur is available for another Zoom meeting the following week. They agree it is time to discuss recent events and the test results at the national laboratory.

McKane is always excited to talk with Matt, who first granted him audience only as a favor to Dr. Pollack. Matt's involvement with McKane has since been as a professor to a protégé. Matt offers advice not as an official representative of the giant corporation; rather, it is his personal interest project.

Their conference is scheduled for an hour but will likely take longer given what is at stake.

"It looks like you have been rubbing shoulders with some very important people—Mr. Bobbitt and Mr. Kirk, impressive. It would be more impressive if your goal was writing science fiction," Matt says.

McKane picks up on the sarcasm.

Matt continues: "You need to figure out the minimum moves to checkmate. The challenge when starting something brand new is convincing the scientific world what you have. You need to continue with your foundation based on simple, scientific validation. You have already taken on the research with PNNL. You have the data. You are getting more data to fortify your research. Your evidence is repeatable. You have two patents. Now is not the time to be swayed by the lure of glamorous people and places. You first need to do something unequivocal in physics."

McKane is listening to every word and knows not to interrupt priceless advice.

"It is not yet obvious what you have done. Tell the scientific world up front the most shocking thing about your discovery. Publish the data in a major scientific journal. Aim high for the most important journals. Why not *Nature* or *Science*? Both are audacious goals. Submit the article and ask for a meeting with the editors. Explain how it will bring prestige to their journal, not depreciate them. If you can land even one of your transmutation claims in your article, that would be big, a career-making event."

Matt adds a warning. "It will be very hard to get the scientists on board. They will criticize your claims. Worse, no one will believe you—especially the claim your device makes excess energy. That is not accepted science. They may not believe you even if it is published."

As if in answer to an unstated question, Matt counsels McKane on how to deal with rejection from the scientific establishment.

"You have a very big idea, but don't force it down anyone's throat. Be the humble inventor. Simply explain this is what you did, and this is what you saw. Be patient with them. Understand they are very smart and well-meaning. When you are talking about your transmutation device, don't preach any scientific theory underpinning the process. Your foundation is data, not concepts. Be careful not to say this device is the "only way" for transmutation to occur because that will offend the sensibilities of the physicists and astrophysicists."

Matt makes a prediction. "If you can get the scientific establishment behind you, you can change the world. If they believe it is really transmutation, then they will not slow walk it. Ultimately, they will follow the best technical talent. They will like your affiliations with PNNL and UW. It is better to have the patents than not. You already have a massive head start."

Matt breaks from his presentation of counsel to McKane to ask him a question related to cold fusion. “Did you know that the Sun cannot produce nickel? So, where do we get it? It likely comes from the fusion power of a supernova. The data indicates your new machine rearranges atomic particles through transmutation. It supposedly forms new elements—including nickel—from basic well water. Do you claim to have a supernova in the chamber of your quantum well?”

McKane answers Matt without hesitation, as if he had pondered the question a hundred times.

“Have you ever considered the power of a supernova can occur at the level of a single electron?”

There is a pause. For a moment, it seems Matt is unable to speak. Then the words come out. “McKane, your brain is way past the Matrix. We are all just trying to catch up.”

A fiery universe of possibilities is spinning around McKane. He senses a planet extending its reach to the stars.

20

Chapter 17: Ambush by the Sea

True to fashion, McKane is first to arrive to the meeting.

He takes in the storied establishment where the investor dinner has been scheduled. A century earlier, the pier supporting the restaurant's structure housed food processing and cannery operations that supplied seafood delicacies to international markets. This venue helped establish Washington State as the salmon fishing capital of the world.

Now the establishment offers fine dining augmented with a prodigious variety of fresh seafood specialties from local waters.

The hostess guides him to a table with an expansive view of the Puget Sound. He thanks her with a smile, mindful to mention her name at least twice during the brief conversation.

That's everyone's favorite sound. Echoes of their own name.

Settling into his seat, McKane notes that it is an unseasonably warm evening for early spring in the Pacific Northwest—perhaps a daily record. It somehow feels even warmer at the table.

McKane removes his dinner jacket, hoping to lighten his mood. Tonight, he wants to set a cool and casual first impression. He's expecting an important guest.

His window seat faces due west with a view opening out onto the sound. The ferry traffic shuttling between Seattle and Bainbridge Island punctuates the quietude of the evening. In the distance, a bold outline of the Island emerges from carbon blue waters. Golden light from the setting sun frames the dramatic ocean-scape. Below the restaurant, boisterous waves dance against massive pilings at high-tide, loudly clapping as if in celebration of winter's end.

Though early, McKane hopes there will be time enough to brace himself for what's about to transpire.

The executive partner of a massive venture capital firm has beckoned him. It will be their first in-person meeting. For months, the Venture Capitalist (the "VC"), whose investment group has nearly a billion dollars to spend on emerging energy sector companies, has been inquiring about McKane's newly patented technology. The VC has hinted at his firm's interest in exploring a formal relationship and has asked for technical data backing up the patents with USPTO and CIPO, including any recent experimental results supporting the transmutation research.

McKane ran an internet search of the VC firm's background and reputation. It has financial roots stemming back to the early 1900s, its wealth based on generations of successful entrepreneurs. Few people would recognize the name of the firm, now a clever acronym. But most would recognize its founders' names, ranking among the richest and most influential dynasties in American history. Now a global company, the present executives troll the world for viable investments.

But behind the firm's storied history lay risks to the entrepreneurs that land on the firm's golden brick road. Rumor has it that once the giant firm acquires a company, their initial concern for the ongoing enterprise and its founders ebbs quickly. And when the acquired company's initial performance drops below par, corporate liquidations soon follow.

Once inside a room, an 800-pound gorilla like this VC firm can stomp you.

McKane calls the waiter over, intending to order a glass of wine. Instead, he finds himself ordering a Jack & Coke with a twist of lime.

McKane has already provided the information requested by the venture capital firm. In fact, he turned the data over without hesitation. He now realizes he should know better.

A wise businessman would slow down, keeping trade secrets close to the vest until the strategic need arises. But McKane is an artist and a scientist—not used to a world of financial sharks. An enthusiastic collaborator and learner, McKane's posture has always been openness and honesty about the technology. Indeed, he is eager to share his novel insights with the world, even more so since adopting a mission to *teach the world*.

If the VC or anyone else wants an explanation on how the magnificent apparatus works, I can do that. Just give me fifteen minutes, he thinks.

Now McKane wonders, less confidently, if the venture capitalist has a new message for him or may even be personally bringing a "term sheet," the first legal document establishing conditions of association between investors and an entrepreneur. What the actual message or purpose of their meeting is, the VC won't say.

McKane only understands that the VC's message is important enough for him to fly in from California.

McKane's guest is twenty minutes late. The inventor wonders if the walk from the downtown hotel room to the waterfront is taking longer than expected. Perhaps his guest is lost along the maze of waterfront buildings.

The delay affords McKane time to think retrospectively about the project. He believes wholeheartedly in the new technology and has held that conviction since the day of its discovery. The present problem is financial. Without an infusion of new capital, there is a threshold beyond which he cannot pass. He has already spent more than a quarter million dollars on institutional research and legal fees establishing intellectual property rights. His savings are nearly exhausted. Funds seemed adequate at the beginning. Now they fall well short of upcoming expenses to advance his work further.

The head teller at the bank has expressed concerns about McKane's ever-dwindling bank account. He tells her honestly about the research and how it may help the world someday. He jokes about the monthly wire transfers to the Department of Energy, suggesting he is doing his part to fund the US Treasury.

The VC finally walks into the restaurant—thirty minutes late.

McKane stands to welcome him. They shake hands. The VC's grip is ferocious. As McKane's guest takes a seat at the table, he apologizes half-heartedly for his tardy appearance.

The waitress returns to serve McKane his drink, just in time. He takes a sip, and then another.

The VC orders his favorite Martini. To McKane it sounds complicated, like a medicinal formula or the "Vesper" version of the Martini from "Casino Royale."

He can't stop himself from imagining that the man across the table is on a secret mission. His guest has brought along a slim black briefcase. Could documents pledging the needed financial commitments be inside?

The VC requests the dinner menu, noting to the waiter that he is only interested in the fresh seafood. Without conferring with his dinner companion, the man orders hors d'oeuvres, a full platter of fresh oysters from Willapa Bay.

McKane picks up on his guest's vibe. The man's tone with the server is laying the groundwork for a take-no-prisoners conversation with the inventor. It strikes McKane that beyond the first impression, he knows next to nothing about the VC.

Conspiracy theories flash through his head.

Could this man be a corporate spy? Is he a covert agent of the fossil fuel industry aiming to undermine the game-changing technology?

McKane lets his rationality prevail. These dark theories have no underpinning evidence.

The VC jumps right into the evening's business, pushing for updates on the quantum research.

McKane dutifully provides the details, touting two additional patent applications in the works, more stunning discoveries in the national laboratory and progress with a journal publication.

Their server delivers the local oysters on an ornate silver platter just as McKane finishes his opening monologue. McKane's guest surveys the food.

"My research article was accepted for review by *NPJ Nature Magazine* in just five days," McKane says.

"I don't believe you. That's impossible. It never happens like that."

"I have the notice right here on my cell phone. You can see it if you like." McKane slides his phone across the table.

The VC steals a glance. "I still don't believe you. It must be some mistake."

McKane registers the insult, a glancing blow.

"Even if true, I doubt the assigned reviewer is even a mainstream scientist," says the VC.

McKane senses a cold, harsh slap to the face. "To the contrary, the reviewer is a worldwide expert. He has been cited in scientific publications over 350 times. Two other reviewers have also been assigned, both with impressive scientific backgrounds. All three scientists express great enthusiasm about the research."

The VC looks unimpressed. Turning away from his dinner companion, the man orders a glass of Washington white wine.

As the server swifts away the empty silver platter, the VC says, "I read the draft of your journal article. It is pure gibberish. I don't understand a word of it."

McKane swallows hard. "Is that your opinion, or do you base it on the review of your consulting PhD physicist?"

"No, I didn't have her read it."

The VC signals no embarrassment for his lack of due diligence. "It doesn't matter, I don't see that you have a product meriting investment or even a scientific review."

Before McKane can respond, the VC interrupts their conversation to order the fresh Halibut. The server stares at McKane.

Unnerved, McKane glances absent-mindedly at the menu. He orders the first item that meets his glance, the Shrimp Tempura.

He tries once again to make some headway with the VC.

"Would you agree that transmutation itself is a game-changing process with applications across countless industries, easily worth billions or even trillions?"

The VC pauses and takes his time to order another glass of white wine.

"That's interesting, but I still don't believe you."

The discussion is going nowhere fast. McKane tries to regroup his thoughts before this potentially promising meeting, and his chance at landing a powerful investor, goes down in flames.

The VC is now sampling an Oregon Merlot. McKane speaks again, this time in a more confident and direct tone.

"We are scheduling another research project over in Richland, at the nuclear facilities. We have already talked with three of the lead scientists there. The suggestion was made that our device might offer a secondary water treatment system to mitigate the release of radioactive wastewater into the Columbia River. If so, it could address impacts from nuclear catastrophes, such as at Fukushima. We also think it may transmute water into tritium on demand. Tritium is worth $30,000 per ounce. It is one of the biggest line items on the Department of Defense's annual budget. While testing, we plan to carefully monitor the extent of excess energy and hydrogen produced by our device. Basically, our apparatus is a pristine energy generator without residual pollution."

The VC sits silently for a few moments. He seems distracted by the first bite of a deliciously prepared halibut. McKane notes how long it takes him to savor the first bites, before finishing with a long sip of Washington white.

The VC looks squarely at McKane, as though prepared to end a battle of wills with one swift stroke.

"Don't go over there to the nuclear facility. Don't do any more testing with the government. They can't help you. It's not the right time."

McKane notices there is no offer of proof as to the meaning of "*the right time.*" An internal alarm in his mind is activated. He's staving off a rush of panic in his brain. *The contest is not over*. His chest tightens. He prepares for the next blow.

The VC delivers it without blinking. "You are not even a scientist. You have no credentials to make your claims or go forward with more research."

McKane sits back in his chair. He could deliver a clever comeback, but instead he deflects with a simple admission. "That's true. I *am* an artist turned inventor. I just happened to make an important scientific discovery."

The VC leans forward. "So, why should anyone believe you? You claim the ability to dissociate any dielectric with your process, transmuting it into new elements. That's impossible. It's a *hoax*. Whatever you are doing, I challenge you to dissociate D2O, the standard dielectric used in cold fusion experiments. No one can dissociate heavy water."

McKane leans in, too.

"I can," he says firmly.

"I don't believe you."

"Give me a week, and I will *show* you."

Backed into a corner, the VC ignores the offer. Instead, he blurts, "What I really want is for you to explain to me how you do transmutation. Give me a simple formula, *A+B=C*. Tell me in a few words and put it down on three-or-four pages max. Send it to me. Then maybe we can talk further about your future."

McKane recognizes the ambush almost too late.

Of course, it's a trap. It only makes sense. Transmutation and cold fusion will be a 40-trillion-dollar industry. There are bound to be competitors out there searching to exploit an opportunity, especially from an inexperienced businessman.

McKane leans back in his seat again. His eyes fall to the nearly untouched plate of food in front of him. He wishes he could be somewhere else, drifting along the z-axis in deep space in his own flying saucer.

Wanting to bring the engagement to a swift end, he contrives a hasty excuse about the lateness of the hour. He stands and thanks the VC for the opportunity. Wishing safe travels to his dinner companion, he exits the restaurant, leaving behind a meal uneaten.

There will be no rematch.

* * *

The sea air is pushing in heavily from the northwest convergence zone, bringing with it cool relief from the heat in the restaurant. McKane's steps to his car are swift and buoyant.

He notes his appetite has suddenly revived.

21

Chapter 18: Vindicated Vision

A lesson learned.

In the wake of the disastrous meeting with the VC, McKane now sees the investment space he's entering for what it really is—full of great opportunities, but also threats. He'll come across potential allies. As well as charlatans.

His most recent guest had come to Seattle under false pretenses, disguised as an inquisitive investor. Perhaps the man was a corporate raider, dangling massive resources just out of reach to entice disclosure of closely held trade secrets and revolutionary experiments.

McKane muses that *VC* could stand for "*Vulture Capital*" just as easily, the kind of firm that circles at a distance until ready to pick up the remains.

It's time to get back to work.

* * *

The first phase of McKane's business plan is complete. Now, McKane must set himself to a crucial task on the trajectory toward success, that of reviewing and refining the journal manuscript submitted to *NPJ Nature Clean Water.*

The renowned publication accepted his scientific article within five days, an amazing accomplishment, and a great sign for the novice inventor. The article has now been assigned to three esteemed scientists for review and editing, each having different technical backgrounds. McKane wonders what the swiftness of the process portends. The future of his work now rests in the hands of this select group of scientists. For McKane to make headway at bringing his invention to market, brilliant scientists must formally recognize the merits of the discovery.

He sits and waits, secreted away in his scientific monastery. While he waits, McKane experiments with the acceleration cycle of his retrofitted, hybrid, hydrogen-fueled BMW engine. He comes to think of the novel engine as powered by the energy of the Sun.

With the balance of his time, he designs another quantum energy experiment, this one even more radical than the last. He predicts the newest experiment will yield the most shocking revelation of all.

* * *

An email notification flashes on his phone. The reviewers' comments from *NPJ Nature* are coming back early.

He hesitates momentarily before opening the message, realizing the significance of this moment. He may be crossing another consequential threshold in his inventor's journey. McKane wonders if the feedback from this publication will further shape or clarify his destiny.

Opening the formal messages, he reads the first round of "Comments to the Author."

Reviewer #1

> *The manuscript on A Novel Resonant Electro-Physical Transmutation Method for Water Purification, by Lee et al., is a unique paper with two patents already published online on water purification technique. The indigenous tri-coil design was designed to overcome the challenge in maintaining the resonance with the dielectric medium during operations. It is extrapolated that it will help scientists to work on a number of exciting new arenas of research where hydrogen production could be of help as*

well as radioactive tritium could be harvested as discussed by the authors. The paper is well thought out with clear objectives integrating logics with experimental proof lending scientific credence to the manuscript. It is an original piece of work undertaken by the authors. The experiments undertaken for analyzing data reach to a reliable conclusion however the chemistry of the processes and their logical discussion is missing.

Reviewer #2:

The manuscript is technically and scientifically sound with alternative intensive studies on water purification using electrochemical methods that is purely physical and devoid of current passage through an electrolyte in the conventional electrolytic method. Average metal concentration of elements pre and post experiment values were determined via ICP-OES. The study had shown that utilization of the resonant magnetic fields in conjunction with electrostatic temporal flux lines aid in dissociation of both FW and SW, with no electrolyte at a very low current density (0.142 mA cm-2) as compared to prior technologies the water purification sector. It also favors the easy removal of toxic metals such as arsenic from fresh water. Going by the convincing results and discussion presented in this manuscript, I recommend its acceptance after attending to the minor corrections.

Reviewer #3:

The authors are commended for the outstanding research presented in the manuscript. The discussions of the findings of this study present a novel means to purify water; a much-needed invention throughout the world. The manuscript is well written however, the authors are requested to read it once more after amending the figures and table indicated above.

McKane is stunned.

He never expected this degree of glowing feedback from all three worldwide experts. He finds himself intrigued by their curiosity on the underlying process of the Quantum Kinetic Fusor.

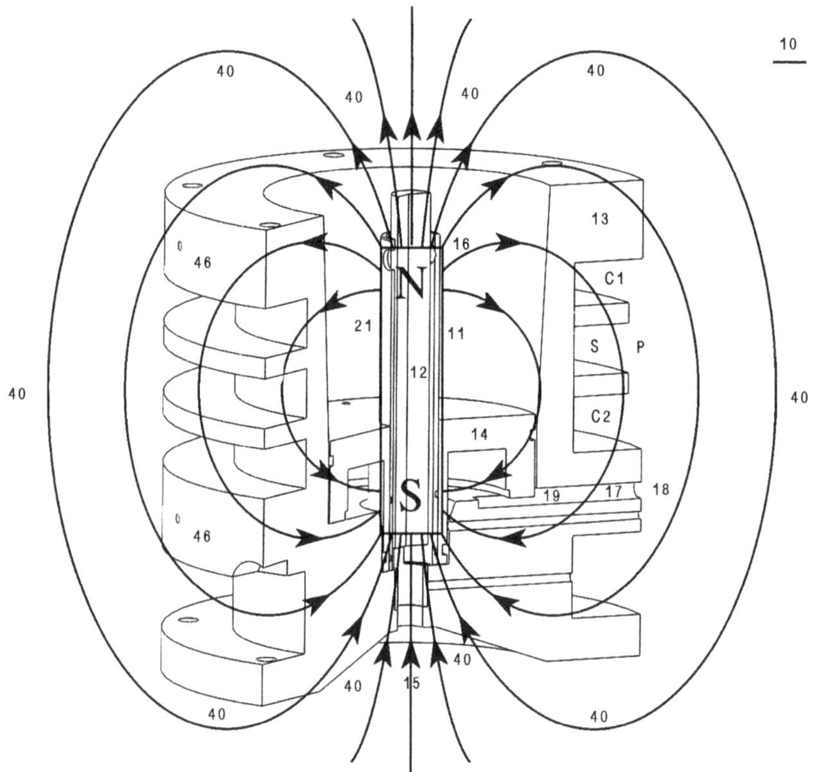

The Quantum Kinetic Fusor

They ask him for more details, please.

McKane has been waiting for this invitation for six years.

22

Chapter 19: Star in a Jar

Going forward, McKane faces a simple but revolutionary objective: the need to establish transmutation, or "synthetic nucleosynthesis," in a way that can't be refuted.

If I can get scientists to understand the reality of nuclear reactions occurring at super low energy inputs in the apparatus, McKane thinks, *then my ideas could change the world.*

And he is convinced that he knows how it can be done.

He must run his experiments again, this time using heavy water (D2O) as the dielectric. He will follow a paradigm established by prior tests that attempted—but failed—to produce cold fusion. In the case of the quantum kinetic well, McKane is confident his results will be different, clearly distinguishing his machines from historical studies.

Interested but skeptical scientists will require him to show how the QKW succeeds in producing repeatable nuclear fusion, where all previous research projects have failed to generate reliable, useful results. If he can demonstrate how the synthetic device can overcome the impossibility of the Coulomb barrier—mother nature's repulsive force field preventing fusion at the atomic level—there's virtually nothing left to stop countless applications of the technology to better the world.

First, though, he must teach the world nuclear fusion as if it were first-year algebra. *A+B=C.*

Despite setbacks in recent months, McKane feels mentally prepared for the challenge. In fact, he can hardly contain his excitement in anticipation of launching the new experiments. In the planning phase, the experimental design for the new trials somehow comes to his mind instantly, as if he had been planning it for years. For once, he senses the momentum of a resonant cresting wave of success just ahead.

That wave is about to break in a secret lab, steps away from the Pacific Ocean.

* * *

The experiment will be done at various energy inputs: 30 volts, 60 volts, 90 volts. Within each phase, run time will be 24 hours in the resonant cavities. With the staggered voltage inputs, the results should reveal whether the QKW effect is linear or exponential. He will use deuterium enriched heavy water this time, extremely expensive but containing double the neutrons of seawater or well water.

McKane wonders in retrospect why he used seawater, well water, and distilled water in all prior experiments. Perhaps he was over-confident in the machine. On the other hand, how magnificent would the outcome be if the new technology worked with basic, cost-effective materials.

Not everyone could afford to spend $5,000 per sample on deuterated water. But anyone could capture water from a domestic well or the seashore.

In McKane's judgment, his prior work—deemed promising but incomplete by the reviewer's standards—was not a waste. It was a revelation. The original experiment established with low power inputs of 1.9 watts or even lower, anyone could reliably produce element transmutation, nuclear signatures, and excess energy.

The problem was that the original water samples, though described accurately by PNNL in pre-test procedures, may be viewed by unconvinced observers as having a level of innate contamination. After all, they will say, isn't ground water "dirty" in some respects, and seawater potentially "contaminated?" What this means is that critical scientific minds might later question the experimental design as well as the results, no matter how remarkable.

The remedy to this conundrum is to use industrially manufactured deuterium enriched heavy water to 99.9% purity. Under these conditions, potential attacks by others seeking to weaken his research findings would be cut off entirely.

* * *

He will conduct the experiments over three days, not long after his cynical guest threw down the gauntlet at their meeting in Seattle.

"I don't believe you."

Those words ring out in McKane's memory, fueling this phase of his work.

He carefully sets up the apparatus and arranges the sensors in place. Next, he positions the electrical connections, requiring only a few minutes. Activating the quantum machine is even more effortless, requiring only turn of a user-friendly electronic switch.

Although the initiation phase of the test is parsimonious, what follows is incomprehensible. Inside the resonant cavity a sudden, complex, quantum kinetic change in atomic isotopes of all selected samples will occur.

McKane now thinks of the process as *a star within a tiny jar—the power of the Sun at the size of an electron.*

* * *

The inventor has never attempted transmutation at only 30 volts. He understands the scientific community will largely dismiss the possibility of a nuclear event occurring at such super low energy input. *That's absurd nonsense,* they will say.

Nevertheless, based on hundreds of trial runs with seawater and well water, he is confident the result of the experiment will prove the same—perhaps far greater in effect—given the density of neutrons in the heavy water.

McKane monitors the machine's operations with a remote camera for the 24-hour periods included in each trial. Three additional instruments will provide measurements for the test. He must take the greatest care to substantiate and document his observations continuously, laboring more than anything else to track the energy levels coursing through the Quantum Well.

In these experiments he is attempting something never done in scientific history: D2O meeting the quantum kinetic well phenomenon. As he begins the trial, McKane again finds himself holding his breath. At the threshold of the unknown he stands again. This time around, he may be constructing a new reality with galactic implications.

But he's come to embrace the flush of excitement that meets him in these moments. To him it's another vaulter's high.

* * *

McKane closely monitors the experimental results. Initial observations are more than intriguing. They are stunning. He can hardly wait for confirmation from the National Laboratory.

* * *

McKane drives across the state to deliver the experimentally treated water and sediment to PNNL. The trek takes place on a Friday morning, immediately following the final trial. He decides to keep the anodes and cathodes for later testing by SEM by a respected laboratory in a small community north of Seattle.

* * *

Nick, the team leader on this test, is there to greet him in the parking lot exactly at 9:00 a.m. as planned. McKane was instructed not to enter the facility itself, since there had been no time to set up the procedures to grant him a security badge. Though unable to visit inside the federal lab, the game-changing tests on the materials from the QK Well can proceed.

McKane catches Nick up on the publication status of his work. Nick can't conceal his surprise that McKane's journal article came to be accepted for review by *NPJ Nature* so swiftly.

"I still can't understand why you got accepted by *NPJ Nature* for review in a matter of a few days. Most scientists that attempt to get published in *Nature* are never accepted over their entire careers. They don't even get to the review stage. Less than 15% get past the first hurdle. You are not a scientist nor an academic, but your article was immediately accepted for review. I have been trying to get published in *Nature* for 10 years and never got a look."

"Beginner's luck," McKane says, hoping to convey empathy to a fellow scientist.

The truth is, he feels both proud and daunted by his success so far.

"I'm keeping my fingers crossed, Nick."

"I will get right on the tests starting today," Nick advises. "We will first have to dilute your water samples one thousand percent, or your treated water will destroy all our instruments. Full analysis may take two months. Stay patient. This is going to be important."

* * *

After leaving PNNL, McKane is first in line at the ferry terminal in Kingston. He will take the next arriving ferry on the 30-minute ride to Edmonds.

As he waits, a police officer with a Homeland Security Badge approaches his vehicle. Her companion is a large police dog, a black Labrador mix. She motions for McKane to roll down the truck window.

"You're not in trouble, and you are not under arrest. I just must make sure my dog gets his treat for the day. Are you carrying a weapon or any ammunition this morning?"

"No ma'am," McKane says.

The Labrador is jumping against the truck's back door, attempting entry through solid steel. Frustrated, the powerful dog crawls under the truck's undercarriage, snorting and whining loudly.

A crowd of curious commuters looks on.

"Well, it seems my dog really likes your truck," says the officer. "Are you sure you don't have a gun or ammunition on board?"

"I don't even own a gun, never have."

With great effort, the officer drags the animal out from beneath McKane's truck. Immediately, the dog slams his head against the truck's back door once again, barking.

The officer appears nonplussed. She poses no further questions and apologizes for her canine companion. Then she moves down the line of vehicles waiting for the ferry.

The ferry arrives, a behemoth floating gently into the terminal stall. McKane eases his F-450 truck into position on the ferry ramp near the head of the line of cars. The ferry peacefully departs. Enjoying the commanding view of the water, McKane reflects on the Homeland Security dog's behavior.

Why would a trained police dog single out his truck as a potential concern for Homeland Security?

Then an idea forms in his mind.

McKane turns and scans the back of his cab. On the backseat are three sealed plastic bags. Inside the bags are the three sets of electrodes used in the recent transmutation experiment. On each electrode is a plating of some deep chocolate brown substance of unknown identity.

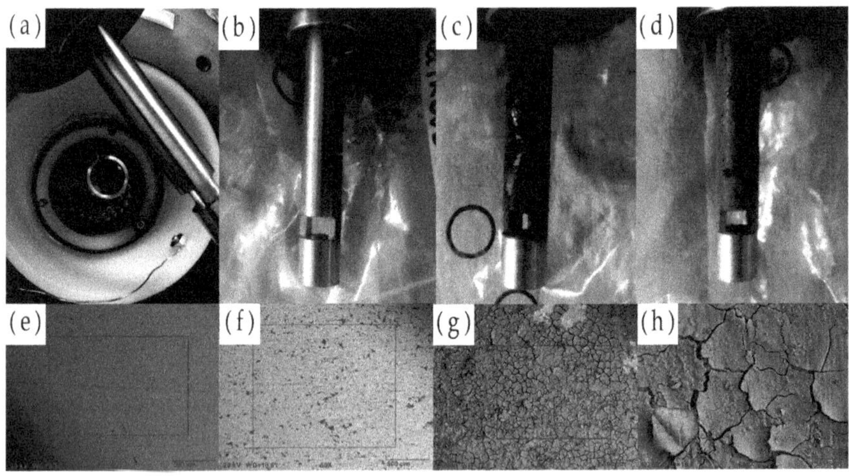

a.) D2O 30v wet b.) D2O 30v dry c.) D2O 60v dry d.) D2O 90v dry

What's in gunpowder?

McKane reaches for his phone and runs a *Google* search.

Moments later he finds that the general components of gunpowder are sulphur, carbon (charcoal), potassium nitrate, and potentially thermite (the only known element that will burn through steel). All are by-products of prior experiments in the quantum kinetic well. Previous experiments consistently yielded sulphur, chlorine, titanium, silver, iron, carbon.

Oh my god, the dog must have detected those elements, he thinks. *That Labrador deserves two treats for the day.*

McKane calls the ferry terminal to tell them. And to apologize.

23

Chapter 20: The Test

McKane's hopes are high for the pending data release from the D2O experiment. Waiting isn't his strong suit, and patience isn't his virtue. He takes solace in Nick's reassuring words, *stay patient—this is going to be important.* Nick forecasts a time-frame of at least two months until the results come through.

* * *

With experiments in the lab at full stop, McKane has time to glance backwards. He reflects on his long-term obsession with nucleosynthesis, the natural nuclear reaction that binds a large atomic nucleus from two or more smaller ones. He remembers well the day its power first manifested in his make-shift laboratory. His initial shock at the instrument readings was soon overwhelmed by a penetrating question, "*Where is all this energy coming from*?"

Little did he understand at the time that he had stimulated a natural nuclear event within the confines of a six-by-six-inch chamber on his bench-top. After years of research, he now understands that the small synthetic device holds the promise of efficiently conquering the Coulomb barrier and thus stimulating nuclear fusion with little electrical effort.

No one had figured it out before. For seventy years, scientists had thought the only way to defeat mother nature's repulsive force field was to smash atomic particles together at super speeds and/or million-degree temperatures. The Sun was their model. After all, the Sun is an immensely hot space body and gigantic nuclear fusion reactor. However, after a generation of colliding particles and mega-million-degree temperatures in sensational facilities, no one could sustain fusion ignition for more than a fraction of a second. All the billions of dollars spent in the effort didn't buy anything of commercial value.

Previously unknown to science was an alternate way to bridle nature's Coulomb barrier. Having his mind on gravity for so long, McKane understands this power to be the only innate process known to de-shield the Coulomb barricade. And when creation's atomic blockade falls, fusion rushes in bringing with it limitless plasma energy.

* * *

As predicted, the lab reports trickle in slowly over a two-month period, and even longer. The unfolding data presents an intriguing pattern of isotopic changes in the dielectric (and even within the electrodes)—a result never seen before in electrolysis-type tests—nor in prior cold fusion experiments.

McKane begins to expect the unexpected with each tranche of new data.

Nick apologizes for delays in finishing the analysis on schedule. Some of the treated D2O samples were analyzed twice. Nick explains the initial findings were so unexpected the National science team was shocked into a state of disbelief. They admitted to initially not believing their own instruments—the best and most capable in the world! And when the team ran the tests a second time, the results were the same! No one had ever seen this before.

* * *

Nick supervises analysis of the post-test samples and reports the data.

McKane composes the narrative summarizing the experimental design, tests, and findings. The paper describes an electron-altering process that stimulates micro-supernova star events like single electron Nucleosynthesis. The resulting physical conditions promote synthetic CNO burning, electron extraction, and room temperature fusion.

Both McKane and Nick understand the publication of the research paper will change the world of science—and human history forever.

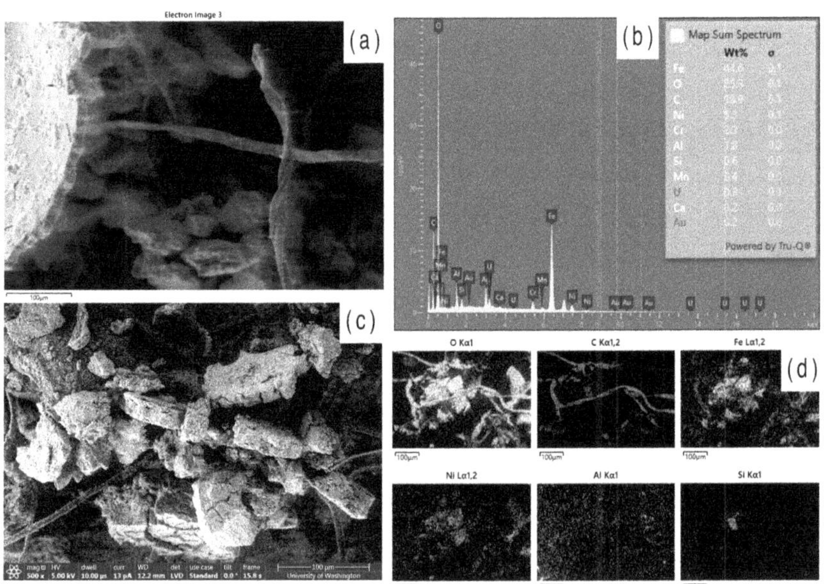

Supplemental Figure 7. Sediment scrubbed off electrodes. **(a)** SEM image of dried sediment. **(b)** EDS values. **(c)** SEM zoom factor 500x. **(d)** MAP readings of sediment. Nano-particles of Al, Si, C, O, Fe, Ni, Au, Ca, and U. The sediment is ferrous and is attracted to a neodymium magnet. Woven shaped Carbon like tentacles detected.

D2O sediment fabrication (Organic & Inorganic)

24

Chapter 21: Censorship

New doorways are opening to McKane. Now that his patents and publications are available for everyone to see, he regularly receives unsolicited inquiries regarding his mysterious technology. Messages roll in from people curious about his inventive machine. *When will it be available?* they impatiently ask.

Other contacts come in the form of offers to explore commercial relationships with technology companies, world banks, and foreign countries. He is asked to write a press release for two major newspapers in the United States. Two documentary filmmakers reach out to him wanting to tell his story to the world. He is invited to be a presenter at an annual climate conference in Europe. A famed tech giant comes out of retirement offering to help him save the environment.

McKane begins to accept he is no longer a recluse living in dismal isolation. No longer an unorthodox artist conducting self-taught experiments the scientific community will never recognize.

* * *

He receives an unexpected message from Roger, his patent lawyer.

Congratulations! I am pleased to inform you that we have now received a Notice of Allowance for another patent application. The rest of the prosecution of the patent application will simply be a matter of carrying through with some administrative formalities on our part. I am so glad for you.

So far, I think we are winning everything. That is due in great measure to the innovative ideas of McKane.

Roger's message draws an immediate response.

To win a race, it helps to get off to a great start. You did that for me, Roger...and I am forever grateful.

P.S. An FBI agent flashed her badge at me several weeks ago. She wants to be friends and meet often. I also had an unsolicited visit from a very pleasant gentleman in a large white van marked "National Intelligence." He stated he wanted to change out my underground internet cable because of a hidden defect.

Seriously, Roger, I am not paranoid - yet!

Best – McKane.

* * *

With so many positive developments stacking up, McKane wonders when his beginner's luck will run out. He recalls warnings from friends about the treacherous territory he is venturing upon for the first time. They explained that as an unpracticed inventor seeking recognition from academia, he would quickly engender opposition.

Fierce opposition.

One of McKane's mentors made specific predictions about how roadblocks would begin to line up between his work and its acceptance by the scientific community. He suggested that academics would try to show him up for ignorance of advanced mathematics and engineering principles. They would further profile him as an over-confident schoolboy, a hack who never completed his studies in the proper major at university. Others would take offense at his unpolished writing style, mocking his whimsical forays into futuristic pie-in-the-sky applications.

These prophetic words of warning echo in McKane's mind.

What you are doing will challenge the stereotype.
Expect people to be incredulous and dismissive of you.
You won't be the first.

* * *

After six years of solo research in his remote laboratory, the headstrong inventor is ready to present a renewable, clean energy platform to the world—and for the first time in history—a functioning Arc Reactor. Some people have told him, including the creator of the movie version of the Arc Reactor, "You are the real Iron Man."

More importantly, his work has birthed clean energy machines like nothing scientists have seen before. One invention cleans contaminated water—even radioactive wastewater—from rivers and oceans. Other devices built by McKane propel water-fueled vehicles with his inspired injector system. But chief among his engineering marvels is the Quantum Kinetic Well (the "QKW™") and the Quantum Kinetic Fusor (the "QKF™"). The two machines coupled together provide a "scalable" working fusion machine.

An achievement previously unknown in Physics.

McKane's eccentric pursuits have led to an unprecedented accomplishment, one that captivates the interest of friends, family, and several key academics. But that is not enough for this scientific maverick. He aims to raise the bar even higher—to be published in a *Nature* journal twice in one year.

McKane takes his passion straight to the front door of scientific publication. He submits his newly completed manuscript, "A Resonant Electro-Physical Transmutation Method for Synthetic Nucleosynthesis," to *Nature*, arguably the top scientific journal in the world.

Though a bold step, the inventor reassures himself by reasoning, *what better way to spread word of these novel methods and gain respectability than from the most vaunted, peer-reviewed publication?*

McKane does not have long to wait before it becomes crystal clear his paper does not stand a chance at the front door, nor even the back door. A week after submission, *Nature* sends out the rejection letter.

Dear Dr. Lee,

Thank you for submitting your manuscript, "A Resonant Electro-Physical Transmutation Method for Synthetic Nucleosynthesis," for consideration. I regret that we are unable to publish it in Nature.

As you may know, we decline a substantial proportion of manuscripts without sending them to referees, so that they may be sent elsewhere without delay. In such cases, even if referees were to certify the manuscript as technically correct, we do not believe that it represents a development of sufficient scientific impact to warrant publication in Nature. These editorial judgements are based on such considerations as the degree of advance provided, the breadth of potential interest to researchers and timeliness.

In this case, we do not feel that your paper has matched our criteria for further consideration. More specifically, while we have no doubt that your ongoing explorations of the unusual physicochemical behaviors exhibited by your 'Quantum Kinetic Fusor' will stimulate others to think about their possible origins, I regret that we are unable to conclude that the paper offers the sort of compelling mechanistic insights that would warrant publication in Nature. We therefore feel that the paper would find a more suitable outlet in another journal.

Please be assured that this editorial decision does not represent a criticism of the quality of your work, nor are we questioning its value to others working in this area. We hope that you will rapidly receive a more favorable response elsewhere.

I am sorry that we cannot respond more positively on this occasion.

Yours sincerely,

Karl

Karl Ziemelis

Physical Sciences Editor, Nature

Swift dismissal strikes McKane like an unexpected punch to the solar plexus. It is even more staggering in view of how other nationally recognized physicists and engineers had previously welcomed the research and his strange new insights into the quantum world. McKane finds his work and writing summarily dismissed without editorial review or even referral to a referee.

McKane shrugs off what could be taken as an insult from a leading publisher. Instead, he thinks about his father's life-long advice to his son, *"Never, ever doubt yourself."* The inventor sends out his manuscript to *Nature Energy* within minutes of receiving the first censorship letter, hoping for a warmer reception.

One week later, he receives a non-acceptance letter from the Senior Editor of *Nature Energy* oddly echoing the same theme from *Nature.*

> *Regretfully, we cannot offer to publish it. Decisions of this kind are made by the editorial staff when it appears that papers, even when technically correct, are unlikely to succeed in the competition for limited space.... We are unable to conclude that the paper provides the sort of significant advance in technological understanding that would be of immediate interest to our broad readership of researchers in the energy community.*

Undaunted, McKane presents the manuscript to *Nature Communications.* This time he is not surprised by the whirlwind brush-off. However, the Associate Editor kindly recommends contacting its sister journal, *Communications Chemistry.*

McKane follows the referral and sends the D2O article to *Nature Communications Chemistry.* His hopes are higher this time, based on the specific referral from the sympathetic reviewer. However, a hurried email soon arrives spurning the research with the same cut-and-paste language as before.

In a last-chance effort, McKane offers the manuscript to *Nature Micro-Gravity.* He anticipates their enthusiastic acceptance of his work since his machine uses quantum gravity waves to stimulate low temperature isotopic fusion. But, the repudiation slip comes with breath-taking speed—in only twenty-four hours!

McKane's profound disappointment at the last message is real but tempered by a new curiosity. He evaluates what just happened in the rapid wake of his submissions to these publications.

Five rejections by notable journals in less than five weeks. Perhaps it is a record—an infamous one. How did the unified messages of dismissal come about?

He tries to fit together the pieces.

How can seven years of grinding research, bold experiments, and dramatic findings be met with wholesale and immediate rejection by orthodox science? Is the current scientific infrastructure a take-it-or-leave-it proposition governed by gatekeepers with the power to dismiss revolutionary ideas and methods out of hand? More importantly, what are the realistic prospects for human evolution if leaders are unwilling to step forward onto uncertain ground?

Upon reflection, the pieces don't fit at all.

Outright rejection from the echelon of scientific society despite doing something unequivocal in Physics–demystifying the enigma of low temperature fusion!

A sadder thought enters his mind, a catastrophic paradox in view of the potential unleashed through his ground-breaking work.

He may be able to stimulate the release of unlimited clean energy from nature itself. *But the world will never benefit from the discovery.*

* * *

McKane returns to his normal working routine in the lab, hoping to shake off the major setbacks to advancing and publicizing his inventions. He resumes work on another electron bifurcating machine—this one with four wheels.

Despite the importance attached to this latest phase of scientific innovation, the inventor finds himself distracted by disappointment and lost in thought. He can't help but dedicate thought to his next move toward an eventual checkmate of his critics and detractors.

The work must be recognized, it's potential tapped by the scientific world before this journey ends. His breakthroughs in quantum physics are not just a fascination, the latest novelty for physicists and chemists. Deep inside McKane is the conviction that these inventions pave a way forward, a way to avoid energy crisis and shore up mankind against some of the weightiest concerns looming on a future horizon.

25

Chapter 22: Touching the Unicorn

Judgment entered in the form of a cold stack of rejections.

An esteemed community of physicists and engineers finds that orthodox physics will not include a low-power atomic manipulating machine capable of converting isotopes into new elements. Further, they find it impossible that a low-voltage transmutation device can convert contaminating isotopes into inert elements.

Hard as it is to believe, for this informal jury of scientists, *seeing is not believing.*

And there can be no appeal—even if the jurists disregard the factual record provided by McKane in his write-up of the results.

Faced with the slamming door of censorship, McKane wonders if going forward means constant contention with an array of influential scientists lined up against him. For now, he fully understands that an organized, academic-based adversary has the power to trample him and quash his case for electro-physical transmutation.

McKane is at another crossroad, despite having found the pearl of great price.

In his favor are three unexpected messages from successful voices in the world of technology and science that just dealt a crippling blow to his confidence.

"Too bad for Nature. Their loss, your gain," states a national expert in the capitalization structures of emerging tech companies.

"The scientific peer review process drives me nuts. It is all a game. But don't worry. There is a better way—having a smarter technology than the other guys." opines the founder of a world-famous technology company.

Nick at PNNL, a close partner in McKane's research, sends a text message of condolence: *"Welcome to my world, McKane. I guess you're not a deity after all."*

* * *

McKane is amused by Nick's heady comment about attainment of demigod status. He sees the double meaning in the message: poking fun at his research partner, while offering an oblique compliment to cheer him up. After two years of diligent research together, McKane now sees Nick as much a friend as a colleague.

McKane soldiers on, stirred by these tokens of timely encouragement. He's embracing a new strategy, to meet directly with a renowned scientist who is skeptical about the technology.

* * *

On the morning he leaves for the peninsula, McKane is thinking about the last time he interacted with Randy. It has been years since their first meeting.

Then, a group of local scientists and educators had assembled to hear a young man present data about a new form of atomic energy. The meeting was Randy's idea. As a university professor for over 30 years, he often took interest in promoting students that showed promise or had specialized skills.

McKane had both in Randy's estimation.

Nevertheless, the meeting itself had been a disaster. Within minutes of beginning his presentation on fusion energy, the enclave of scientists ripped into McKane, insulting him with inane questions and disparaging remarks. When one of the attendees shouted out "bullshit," Randy mercifully intervened, dismissing the hecklers with a fiery look and a wave of his hand.

Since that calamitous day, McKane saw Randy as a heroic figure. A professor, lofty but easy to know. A forensic scientist, legendary for solving complex puzzles. A public servant, sought out by state and federal government for consultation. A family man, gentle yet strong. A friend, affable yet always dignified.

* * *

McKane has been on the road for an hour after disembarking the ferry. He is traveling a scenic route along the Puget Sound, the most popular passageway to the western boundaries of the Olympic Peninsula.

Outside, the sky is engulfed in white light from a luminous sunrise. The ocean's edge is carbon blue and boundless. Seabirds meriting art prints by Audubon populate the shoreline.

McKane's destination is an abandoned warehouse near a remote fishing village. The rickety old building had long been used for boat storage by a once thriving fishing fleet. But today, the monstrous relic stands as a dismal reminder of an industry plagued by misfortune and resource devastation.

As the Ford 450 rounds the final bend just before the marina, McKane sees the massive structure perched alone atop a barren headland. He eases into a vacant, unmarked parking area, covered with weeds and tall field grasses. On the building is a large sign: "Under New Management."

Beyond that, there is no sign of life to be seen here.

The front door is locked, but McKane has the key. Randy entrusted it to him for ready access whenever needed. The professor purchased this property to serve as an adjunct workshop for his fabrication hobbies and as winter storage for his fishing boat. There was also the obvious future commercial opportunity to renovate the building into a rustic bed and breakfast.

The high bank location commands views of the Strait of Juan de Fuca and the majestic San Juan Islands to the north.

Since that rowdy meeting with the dismissive scientists, Randy had followed with interest his protégé's struggles to gain public recognition for his inventive machines. Although the two of them communicated on a regular basis, Randy had never fully grasped the underlying basis of McKane's ostentatious claim to produce nuclear reactions at only 1.9 watts of electrical input. The premise seemed intriguing, but improbable.

Still, the professor was careful not to distance himself from McKane's ongoing research and scientific publications.

McKane's purpose in visiting the warehouse today is to address his mentor's lingering doubts head-on. It is a rendezvous of two respectful and supportive colleagues. But in McKane's gut this meeting feels more like a showdown. The inventor has implicit trust in his machine to replicate its mysterious process anywhere and at any time. Even at an abandoned boat house.

The question is whether he can win over the incredulous professor's full confidence before he leaves.

* * *

Randy and McKane are in the boat-garage-turned-makeshift-laboratory for the day, both staring at the Quantum Kinetic Fusor, which stands unpacked and fully prepped on a timeworn workbench.

Although the professor has read the patents and the journal articles, he has never examined the device in a forensic setting. McKane aims to demonstrate the quantum machine in action while the professor watches.

Randy approaches the workbench to examine the setup.

"It's ready now," McKane says. "It's locked into resonance with the water. All you do to activate the process is turn the voltage knob." McKane looks at the professor. "Do you want to turn the knob?"

Randy hesitates. He did not expect direct involvement in the experiment. In fact, he expected the opposite since his duty as a scientist limits his active participation in a testing regime. Typically, he would establish experimental protocols for the test, then carefully observe the machine under study as a scientific witness.

But this case is different.

McKane is more than a client, more than a respected junior scientist. He is Randy's friend.

So, Randy accepts the challenge. He cautiously turns the voltage knob up.

From 0 volts to 10 volts.

From 10 volts to 20 volts.

From 20 volts to 30 volts.

From 30 volts to 40 volts.

Suddenly, Randy stops as if struck in the face by an invisible fist. He lets go of the attenuation knob and staggers back several feet. He leans against the opposite workbench.

"McKane, I have to tell you something," he says in a shaken voice. "I have been around a lot of powerful machines over the years. Some operate by huge electrical force. Some operate on nuclear power. I 'm not one to get physical reactions around machines. Matter of fact, I've only had two headaches in my entire life that I can remember. Right now, I have a headache."

Randy eases himself down to a sitting position on a battered old chair. He pauses a moment before asking a question.

"It's from the machine, isn't it?"

McKane reaches toward the machine and returns the voltage knob back to zero. Then he turns towards Randy and says, "Welcome to my universe. Now you know what it feels like to step inside a gravity well."

"I never expected this," the professor admits. "This power is amazing."

26

Chapter 23: Race to Fusion

"It is going to be a race," Matt said.

"The first one to put a timestamp on the discovery of fusion wins in the minds of the public. That's all important. It's the whole ballgame."

Matt's prophetic words were spoken in their first zoom conference. His warning was unmistakable and ominous.

But there was more. Matt explained what "winning the race" means.

"You already have the data, everything we are looking for. The challenge is to convince the world what you have. Tell us what you have done. Show the world you have done something unequivocal in physics. Open Pandora's box. Put your science on the table. Publish your work in an important scientific journal. Confirm your discovery with simple scientific validation. Establish your economic propositions. Explain why you should be the one telling the story."

* * *

On December 13, 2022, McKane hears shocking news. In a surprise announcement during a press conference, US Secretary of Energy Jennifer Granholm proclaimed a major scientific breakthrough in fusion research at the Lawrence Livermore National Laboratory ("LLNL") in California. According to Federal officials, LLNL researchers had successfully ignited a first-ever fusion reaction that propagated more energy than it took to begin the process.

News concerning the successful experiment is flooding the internet and dominating the leads on prime-time cable. After 70 years of research at Lawrence Livermore, the improbable has finally happened.

Official reports indicate the atom-smashing demonstration employed the largest array of lasers on earth, 192 in all, housed within a building three times the size of a football field, to blast a spec of deuterium until it exploded inside a containment vessel. Although the fusion event only lasted a billionth of a second, the accomplishment was being hailed as a major "breakthrough" and a "landmark achievement."

McKane thinks the breaking news impossible.

He sits down on a lab chair next to his library of old books. In need of distraction for a minute or two, McKane steals an upward glance as if to acknowledge these old companions—the books he has read and reread over a decade. Iconic books about gravity and electromagnetism. Biographies of James Clerk Maxwell, Oliver Heaviside, and Nikola Tesla.

Masterminds whose ideas changed the world.

Today, an astonished McKane wishes for inspiration from the animal spirits of these scientific mavericks of the past.

* * *

McKane is conflicted about the Lawrence Livermore news. He feels an initial rush of excitement over the apparent significance of the transformational experiment, portending good news for the planet and its inhabitants.

But he is also devastated. After all his effort and sacrifice, he will not be the first to place a timestamp on fusion.

He lost the race.

McKane scolds himself for not going public sooner when he first had his fusion invention in hand. He wonders why he had been so cautious by insisting on years of arduous research before breaking the good news in a worldwide press release. The delay cost him the opportunity to write his name in a mythical physics hall of fame. More importantly, the postponement cost precious years in the world's fight against the catastrophes of a warming planet.

The cascade of second-guesses forming in his mind deepens his unshakeable feeling of disappointment.

McKane closes his eyes, pausing the moment. He centers his thoughts on steady breathing, hoping to avoid descending the slippery slope of self-doubt.

When he opens his eyes, a sudden realization hits him.

No way! Too much diligence is the right amount of care before making such a tantalizing announcement to a waiting world. Wisdom dictates it better to be the stealthy tortoise rather than the sprinting hare, when chances are the public will freak out by the implications of such a far-reaching proclamation.

McKane is certain that his plodding methodology had been the right call all along. He had opened fusion's Pandora's box one inch at a time.

By first doing something unequivocal in physics.

By confirming the repeatability of transmutation at low voltage.

By establishing his intellectual property rights internationally.

By validating his discovery with scientific research at a national laboratory.

By publishing the dramatic experimental findings in a notable journal.

By revealing the vast commercial applications of fusion.

By extending fusion to the world now, rather than the distant future.

Despite these reassurances that he had charted the appropriate path toward ground-breaking innovation in fusion technology, McKane agonizes over the ultimate question: *Who should be telling the story of fusion?*

* * *

McKane needs a plan of action to get the ground-breaking news of his work outside of his inventor's bunker. To take his work public.

He must find the most adept person in the world who is currently writing about fusion technology. He must reach out to the best global data service company that works with IP owners. He must forge relationships with media empires to disseminate his story. He must contact the bipartisan U.S. Senate's fusion energy committee. He must dine with the governor.

It is time for a Press Release.

Stealth Laboratory Beats Livermore to Sustained Plasma Fusion Ignition

SEATTLE, Date, 2023, Quantum Kinetics Corporation, a company focused on becoming a leader in fusion technology, emerged from stealth mode in response to the U.S. Department of Energy (DOE) announcement of the achievement of fusion ignition at Lawrence Livermore National Laboratory ("LLNL").

Quantum Kinetics Corporation (QKC) is making public, for the first time, the existence of a modular, synthetic device capable of producing atomic fusion at 1.9 – 4.4 watts input. The patented invention, as described by laboratory officials, has the potential to provide limitless, carbon-free electricity to the world – to be used to fuel and heal our planet.

Straight from the realms of science fiction, comic book tales and blockbuster movies, QKC has invented a real-life, functioning Arc Reactor - capable of producing controlled, scaled, and sustained nuclear fusion at room temperatures.

Inspired by a young American scientist who had been studying gravity and hydrogen densities in his garage for 11 years, QKC has successfully demonstrated safe and maintainable fusion – a technology that could permanently alter the course of human history. Executives of the company call the new process "Electro-Physical Transmutation" (E-PT™). The method works with distilled water, freshwater, seawater, and heavy water (D2O) dielectrics as its fuel source. E-PT™ operations can be fully vectored to meet any desired work requirements.

The company recently conducted research on its patented technology in a joint-strategic relationship with Pacific Northwest National Laboratories, a division of the U.S. Department of Energy. The research was published in *NPJ Nature* and heralded by the three journal reviewers as a "*much needed invention around the world.*"

For more than sixty years, nuclear fusion projects around the world have pursued the goal of producing abundant, clean energy by heating matter to more than a hundred million degrees until the super-heated atoms melt into a magnetically confined plasma field, thus replicating conditions in the center of a star. This is the "star building" model to create nuclear fusion. Despite investments of billions of dollars into research and development, the goal of producing sustainable stellar-like fusion remains as-yet unsolved due to overwhelming complications and technical challenges.

Recently, December 5th, 2022, scientists at the Lawrence Livermore National Laboratory ("LLNL") in California hailed a successful fusion "ignition" experiment producing net energy gain. The demonstration employed 192 of the largest lasers on earth, housed within a building three times the size of a football field, to blast a spec of deuterium until it exploded inside a containment vessel. According to Federal scientists, the one-time event was a major "breakthrough" and a "landmark milestone achievement."

Despite the significance of the discovery, some scientists at Lawrence Livermore National Laboratory, including researcher and spokesman Alex Zylstra, cautioned that practical applications of fusion power remain a distant pursuit. Best estimates are that it may still take 20-50 years to realize fusion's commercial propositions as demonstrated at LLNL.

While LLNL was aggressively making strides of stellar-like fusion, a small U.S. energy company was cracking the low temperature fusion code -- bringing the idea out of the realm of myth and into the material world of capital markets and scientific measures – into a space where it can be used for the good of mankind and the earth we populate.

About Quantum Kinetics Corporation (QKC)

QKC holds patents both nationally and internationally on fusion devices. Company officials assess its machine's capability to produce energy gains over losses is attributed to low voltage synthesis of heavy atomic nano-elements. The ET™ process is also distinguished by Biological Index Stimulation of unique diatoms and life structures previously unknown to science.

Recent economic studies estimate nuclear fusion to be a multi-trillion-dollar industry. As such, it is projected to be one of the largest, if not *the* largest, industry in the world.

QKC projects its revolutionary work with fusion energy will provide for the world's poor and under-resourced. And it will open countless doors into a future touched by new hope and healing – and someday a gateway to the stars.

For Further Information visit: http://www.quantumkinetics.co

* * *

McKane is convinced that from now on he must take control of the story. There's no more time to spare. Getting an article published in a prestigious journal, coupled with a national press release, is a great start. But commercialization of an unprecedented scientific discovery is only bolstered by massive publicity.

His invention needs a wider audience. And a brand.

McKane decides to up the ante by creating a "Quantum Kinetic Company" website and an accompanying YouTube video to take his breakthrough technology worldwide. The time has come to enter the larger conversation and put his invention before the eyes of the public.

He understands the consequences of stepping onto the public stage all too well. Loss of anonymity. A storm of controversy in the academic community. Public quarrels about scientific validity.

He expects a barrage of thinly veiled accusations on social media that his emergent technology is just another phony 'get rich' scheme. And he's already bracing for the widespread insinuations from the energy sectors that his story is 'too good to be true.' Stalkers and detractors will come out of the woodwork the moment he pushes his work into the public eye. It's to be expected, and these are serious costs to consider.

However, McKane addresses his own internal Socratic debate about this public strategy with a non sequitur. *The essence of 'cold fusion' is technology so amazing it initially startles the intellect. To doubt such news at first is a natural response. But people will eventually warm up to the idea. They simply need time.*

And McKane has none left to spare.

Convinced of this fact, he jumps headlong into the fray.

27

Chapter 24: Friends Into Family

McKane's background as an athlete helped form his vision of the role of teammates in achieving personal success. Stirred on by teammates, the gymnast who falls off the rings gets up and tries again until the skill becomes second nature. And when the pole-vaulter takes the inevitable hard plummet from a height above 18 feet, teammates press even harder.

Victory, in short, is never achieved alone.

Now, on the precipice of unveiling his breakthrough inventions to the world, McKane leans on the constant encouragement from his team of supporters. As an inventor, he's re-learning the athletes' creed: *The flip-side of every failure is an inspiration to keep trying.*

* * *

McKane remembers the drive to Kerri's house.

It is a June afternoon, and it is sunny. The route takes him through an affluent neighborhood with upscale homes on expansive, manicured acreages. He is enjoying the scenery and leisurely drive in his classic 32-year-old, hydrogen-hybrid BMW. The car still gets its share of thumbs up from other drivers. He welcomes the attention as if each wave signifies the universal acknowledgement that *they don't make them the way they used to.*

He doesn't know how many guests will be attending Kerri's dinner party.

Some friends will be there, along with a few executive retirees from Boeing. He has never met these former administrators of the aeronautical giant. Butler, his honorary grandfather, will be there too. In fact, the party was Butler's idea, another opportunity for his adopted grandson to spread the word about a new form of plasma energy.

It is going to be the first time McKane meets with a group of potential investors. He wonders if the meeting will bring any serious attention to his technology company. He thinks about making a formal presentation to the group but quickly dismisses the notion.

All he wants to do is enjoy an evening with supportive friends and savor a wonderful meal prepared by Kerri herself. It should be something special – after all she is the owner of several fine dining establishments in the Pacific Northwest.

He arrives 30 minutes late but is certain Kerri won't mind, believing she intends their meeting to be a casual and open-ended affair. He pulls his "Bimmer" into a parking spot next a new M car. He wonders who owns the European Autobahn-inspired race car.

McKane is not used to arriving late. He makes a courteous apology to Kerri at the front door, who waves it off as unnecessary. Without hesitation or comment, she ushers him into an open and appealing patio area. Other guests are already assembled around a large dining table. Surrounding the main table are smaller tables offering an abundance of hors d'oeuvres and open bottles of imported Italian wine. The engraved wine glasses are neatly positioned at each place setting.

He sees there are only two remaining empty seats. Both are at the center of the dining table. Adjacent seats.

He takes one of the remaining chairs. Moments later, the final guest arrives as if timed perfectly by the hostess. He takes the seat next to McKane.

Butler makes the introduction. His name is Bob. He spent a long and successful career as an aeronautical engineer at the Boeing Company.

He has seen it all. Done it all. From start to finish, he knows how to get big things off the ground and flying.

Butler coyly asks Bob if he read the journal article sent to him earlier in the week, the paper about "Electro-Physical Transmutation"?

Bob smiles. "Of course, I did. It is fascinating. There are so many questions to ask the author. Whoever wrote it must be quite intelligent."

Butler pauses, sensing the cast of characters is now properly assembled and the audience quietly expectant. With perfect timing, he makes the dramatic revelation as if announcing, *curtains up!*

"Lucky for you the author is here with us today. He is the gentleman sitting next to you!"

The surprise announcement draws an exclamation from Bob. "OMG. He is so young! I would never have suspected that!"

* * *

Bob asks his questions for the next two hours. From center stage, McKane does his best to answer. The discussion is far-ranging. It captivates the larger audience listening in. Soon, Bob poses a direct and unexpected question, "How much money do you need?"

* * *

One of the guests offers to take McKane for a ride in his M car before he leaves the party. Overhearing the offer, Bob follows them to the parking area. But there is only room for two in the race car. Bob grabs the driver's arm and whispers in his ear.

"Be careful. Your passenger is the most important person in the world."

28

Chapter 25: What Many Hands Make

McKane did the impossible – stimulate something living out of something lifeless. By making the first real-life and fully functional "Genesis Device," he proved Lord Byron's 200-year-old proverb to be true.

Life is stranger than fiction.

And there's more. His amazing machine crosses the impassable barrier between inorganic and organic chemistry, while at the same time releasing unlimited power from nature's own free-energy domain. Despite accomplishing the improbable, McKane worries,

People will say it is too good to be true.

Yet, the research shows each D2O test produces organic carbon from 99.9% pure, inorganic deuterated water. The CDOM non-filterable organic carbon increased 1,500%, according to the National laboratory. Worm-like structures appeared spontaneously and then transitioned through complete life cycles within 24 hours. The non-filterable organic material appeared in each test, with the 30-volt treatment stimulating the largest population of organic carbon.

With each tranche of confirming evidence, the entire world is becoming richer, revitalized. But only if

Seeing is believing.

McKane postulates the staged 30-60-90-volt treatments captured element fusion at various transitional stages of transmutation. The evidence shows the volume of fused substances is related to the intensity of the applied voltage.

He sees the E-PT*TM* process akin to organic alchemy coupled with inorganic metallurgy. No wonder the national science team was taken aback at their initial analysis of post-treatment D2O samples. They ran re-tests to confirm the findings. Now the team is standing strong behind the data and publishing the results on the DOE and PNNL websites.

McKane did not accomplish the impossible alone.

* * *

McKane considers the long cast of supporters, antagonists, skeptics, partners, and friends who played a role in his improbable journey as a lost artist, a broken firefighter, a desperate dreamer, and a revolutionary inventor.

He thanks God most of all for his father, Riley—McKane's greatest supporter.

And for Butler, Matt, Nick, Roger Emerson, Dr. Pollack, Randy, and Scott Braswell among others. For the kind widow who rented him the garage where his invention was born.

So many voices injecting hope into a bleak world. Animating his work. Lighting his path. Propelling him to success.

And obstacles, too. Cynics lined up against him. Odds stacked the wrong way at so many junctures. Heartbreaks built on the ruins of other heartbreaks.

And yet they form and shape his story, like pressure and heat bringing diamonds out of coal.

Somehow, he's grateful for these, too.

McKane finds himself on the brink of a fulfillment of a dream. A product of heartbreak and healing. Aiming where he has no business to aim and dreaming something he has no business to dream.

Somehow, he's cracked the code. Brought cold fusion out of the realm of myth and into the material world of capital markets and scientific measures—into a space where it can be used to heal the planet.

His work will provide for the poor and under-resourced. It will open countless doors into a future touched by new hope and healing.

It took an outsider to crack that code. An unconventional coach and athlete. A student dismissed by engineering classmates and industry scientists for what they deemed a lack of respectable credentials.

Nevertheless, he cracked the code humanity was unwittingly waiting for someone to step forward and solve.

He pulled the sword out of the enchanted stone.

And all it took was an artist.

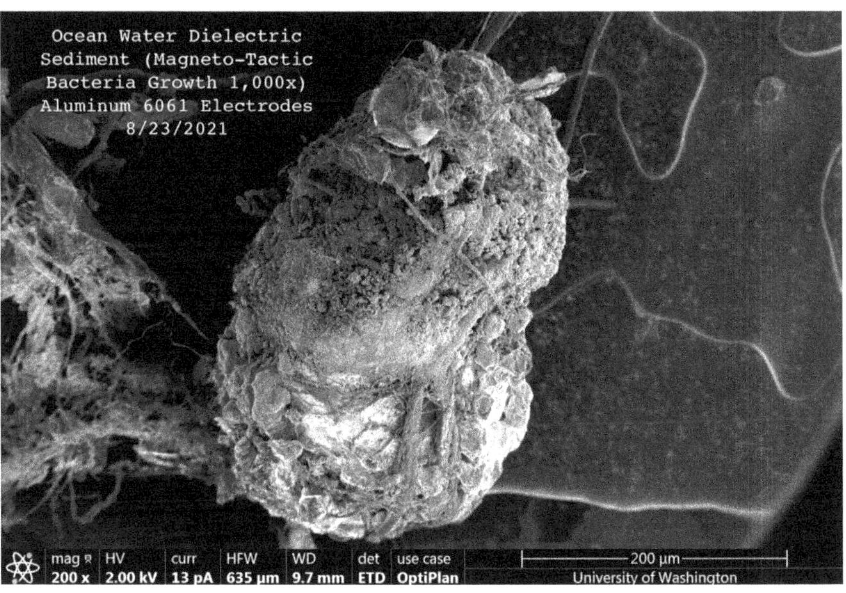

Magneto-tactic Bacteria 1,000x growth post treatment 8hrs

29

Chapter 26: At the Edge of Paradise

It is the third Friday of February. Outside the lab, temperatures have fallen below 20 degrees for the third day in a row. A blanket of ice and snow is thickening by the minute. Soon, McKane will be landlocked on a steep hill with outside temperatures that could kill an unprepared man in minutes.

With a glance at the treacherous snowscape outside, his mind slips back to the day of the accident.

The Subaru revolving in mid-air. The desperate screams. The crawl of each moment as he hung helpless between life and death . . .

McKane has not allowed himself to remember Happy for years. At least not as anything more than a concept. It seems a lifetime ago, before the fateful road trip along a treacherous mountain corridor. But this frigid and lonely morning is different.

They had been married for two years and still blissfully lost in the wonder of an improbable romance. Their home was on Sioux Drive—a fitting address, given McKane's ancestral Indian bloodline. One day he hoped to pass that ancestry on his mother's side to children.

To his and Happy's children.

He reflects on their initial plans for a family together, the prime purpose for their union. He recalls an old and cold decision. One that still haunts him from time to time.

Perhaps it wasn't too soon for children after all.

He can still hear the echoes of Happy's tired admonitions, scolding him for spending too much time on his experimental, hydrogen-powered car out in the garage. But she espoused a probability he could never accept.

It's not going to work.

* * *

An incoming email grabs McKane's attention—an important message from Ohio is coming through. It is Roger Emerson, confirming the acceptance by the USPTO of his "Quantum Injector" patent—the water-based fuel technology to revolutionize and retrofit internal combustion engines everywhere.

With this patent, the game has officially changed for automobile technology. For his invention generates explosive energy instantly from simple water mist. It is the mythical "repulsor."

Water Mist Injector (Repulsor Technology)

McKane is more than astonished by the swiftness of the award, only eight months. He had secretly wondered if the Patent Office would even endorse his novel energy technology—a source of power surpassing both gasoline and diesel.

More readily available in Nature's storehouse.

More economical to serve the disenfranchised of the world.

Potent as a tool to destroy the specter of carbon build-up in the world's atmosphere.

He will call it "Star in a Car." But the applications are limitless. The tech can be retrofitted into any internal combustion engine, including massive machines and even jet planes.

McKane can't come out of the garage now—not even for a minute.

* * *

McKane is on a two-week vacation in Kauai. His father, the psychologist, had finally persuaded him to step out of his inventor's bunker. It would be a "mental health" holiday, explained his father—time enough to recharge his creative energy reserves with long hikes into remote landscapes, endless beaches, and waterfalls.

The barefoot existence would be the perfect counter measure to the loneliness of the lab, the pressure of verifying a new technology, and the burden of sharing it with the world.

* * *

He is on a long hike that lasts the day. On the return leg, he passes a young lady sitting alone on a ledge overlooking the ocean. She appears to be writing something in a small book. They acknowledge each other with brief "hellos." As he takes the next turn in the trail, he stops in his tracks. He suddenly remembers the terrible things that can happen on the edge of a cliff.

Life-ending things.

He quickly starts back, hoping to find the young lady still perched at her lookout. When he arrives, she is still writing.

It is an awkward moment. He realizes she may feel threatened and race away. He quickly tries to explain himself. "Hello again. Sorry, but I just walked past the most beautiful woman in the world. I had to return to tell you."

The young woman smiles, amused by the transparent flattery from a handsome stranger. It is clear she will not be fleeing on the reverse footpath. Instead, she introduces herself as "Sommer." She explains she is a doctor on the island. She often comes to the ocean observation post to write in her diary.

She is curious about him. “Are you a visitor or an islander?”

“I’m sorry to say just a tourist,” he replies. “Maybe someday that will change.”

“What do you do for work?” she asks.

“I do research on water and energy.”

Pointing to the ocean below and to an endless blue horizon, he says, “I can turn a cup of seawater into a synthetic meteoroid, then use the process to generate endless clean energy.”

Amazed or in disbelief, she asks, “How can you do that?”

“It’s quite a story,” he replies.

END

At the Edge of Paradise

www.ingramcontent.com/pod-product-compliance
Ingram Content Group UK Ltd.
Pitfield, Milton Keynes, MK11 3LW, UK
UKHW022002270726
14060UKWH00007B/740/J

9 798218 180508